SPACE ODYSSEY

The first forty years of space exploration

Serge Brunier

SPACE ODYSSEY

The first forty years of space exploration

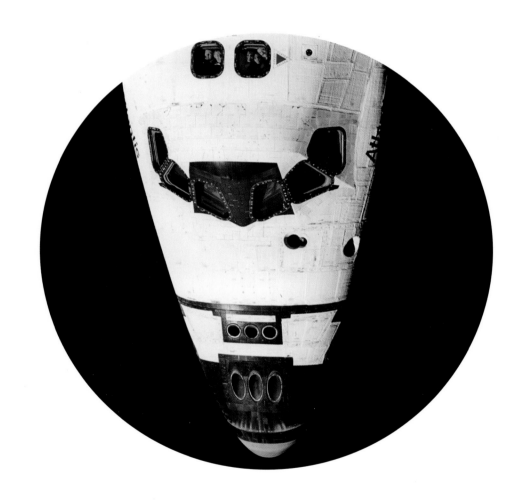

Translated by Stephen Lyle

ACKNOWLEDGEMENTS

The author extends his warmest thanks to Marie-Pierre Levallois and Catherine Delprat for their unfailing confidence, Sylvie Cattaneo and Valérie Perrin for their original and refreshing view of the sky, and Gilles Seegmuller for his long-standing and invaluable fellowship and the individuality he is able to bring to books. Thanks also go to Amandine Six, Julien Guillaume, Jean-Pierre Casamayou and Christian Lardier for their friendly and decisive assistance, Stéphane Aubin for his help and patience, Jean-François Robredo for so opportunely reminding him that the cosmos was Greek before it became Russian, and Christine for her generosity and encouragement. In particular, the author would like to thank all those men and women who venture into space.

This book is dedicated to Michèle, Vincent and Julien Defait and to the memory of Jean-Pierre Defait who has remained so present throughout this adventure.

PUBLISHED BY THE PRESS SYNDICATE OF THE UNIVERSITY OF CAMBRIDGE
The Pitt Building, Trumpington Street, Cambridge, United Kingdom

CAMBRIDGE UNIVERSITY PRESS
The Edinburgh Building, Cambridge CB2 2RU, UK
40 West 20th Street, New York NY 10011-4211, USA
477 Williamstown Road, Port Melbourne, Vic 3207, Australia
Ruiz de Alarcón 13, 28014 Madrid, Spain
Dock House, The Waterfront, Cape Town 8001, South Africa

http://www.cambridge.org

French edition © Bordas/ H.E.R., 2000
English translation © Cambridge University Press 2002

First published in French as *Une Odyssée de l'espace*, by Serge Brunier 2000
First English publication 2002

Printed in France by Pollina s.a., Luçon 85400 - N° L86021

A catalogue record for this book is available from the British Library

ISBN 0 521 81356 5 hardback

CONTENTS

PREFACE

It was a beautiful afternoon in July 1994, deep in the south of France, in the magnificent wild site known as the *Théâtre de la Pleine Lune* not far from the village of Gourgoubès. In the presence of astrophysicist Hubert Reeves and newsreader Claude Sérillon, a small group of astronomers were preparing for the documentary programme *La Nuit des Étoiles* for French television. I was there as a guest, enjoying this peaceful locality with Jean-Pierre Haigneré. He recalled his first flight aboard the space station Mir as we admired several hang-gliders wheeling high over the dry plateau. And then the guest of honour arrived. Not speaking a word of French, he had just flown in from Washington and was rather tired. The evening promised to be long and he wished to rest before the questions began. The same questions as he had been asked now for almost exactly twenty-five years. I offered to be his mentor until the evening began and hence found myself in his company for a few hours. I explained how the *Théâtre de la Pleine Lune* would be brought to life for the occasion and he was remarking upon the beauty of the summer sky, when suddenly the Moon went behind a cloud, leaving the heavens open for the stars and planets. We recalled the unexpected comet which, after a year bearing down upon Jupiter, had finally struck its target in a fine show of cosmic fireworks a few days earlier.

Looking at him, I could hardly believe these magical moments were real, savouring each instant spent in his presence. I imagined him inside his minuscule capsule with the two other astronauts, lodged right at the top of the gigantic rocket. I saw him during his long cosmic crossing and especially at the critical moment when his vehicle had almost used up all its propellant, one finger on the emergency recovery button, wondering for a split second whether his incredible odyssey would come to a sudden and brutal end in a cloud of dust and grey ash. That evening, when the Moon had risen, its dark seas sketching out a familiar contemplative expression, I could not help wondering how he viewed our heavenly companion. Then, just as we reached the television set, a young amateur astronomer asked him in English: 'Edwin, would you sign this photograph?' In a dry and irritated, even aggressive tone, the American astronaut replied: 'I'm not Edwin. My name is Buzz, Buzz Aldrin.' The return to Earth had been difficult for the first man to walk on the Moon, with Neil Armstrong, twenty-five years before. The Moon had left a bitter taste. Whereas Neil Armstrong had chosen to live almost twenty years in anonymity and silence, Aldrin had set out to change his life in every way, even going so far as to change his first name. To appease their passions, these exceptionally highly trained men had taken quite extravagant risks, reacting to dangerous and complex situations with uncommon self-control. They had walked upon another world. And yet, neither superbeings nor robots, they would reveal all their frailty and humanity in the years to follow the Homeric adventure they had helped to write.

One reason put forward publicly to explain why Neil Armstrong and Buzz Aldrin had retained such bitterness after their marvellous cosmic adventure was that both firmly believed they were going to the Moon on a reconnaissance mission. They saw themselves as scouts boldly marking out the way for future night trains to the stars, via the Moon, Mars or Titan.

But no. The Apollo programme was discreetly brought to a close, before its day, and since then no one has reached the second cosmic velocity, the velocity required to escape from Earth's gravitational pull. For thirty years now, astronauts have been flying round the Earth in circles. Like Gagarin before them, transformed into an idol by the Soviet people, the living symbols of space conquest, Armstrong and Aldrin never again returned to space.

On this moonlit summer evening, Buzz Aldrin wore a mixed expression of enthusiasm and disillusion, the old emotion still alive – a sweet bitterness close to bliss, as Buzzati might have said. And then I understood something about the conquest of space, this technoscientific saga that had swallowed up hundreds of billions of dollars, mobilising laboratories, research centres and industry in dozens of countries around the world. I understood that this strange cosmic dream with its near-military organisation, blending together geopolitical intrigue, scientific and metaphysical ideals, and mundane questions of finance, was indeed above all other things a human adventure.

Today, in the year 2002, four hundred men and women have been into space. These astronauts share the same sweet melancholy and yearning as Aldrin and Armstrong when they were so harshly brought back to Earth. With the dawn of the new millennium, they may reckon the progress made toward the stars, whilst counting off the faded Utopias. Between dreams of strolling in the Earthshine and hopes of setting foot upon Mars, they may well wonder whether they are the very last heroes of the twentieth century or the vanguard of the burgeoning stellar form of humanity first imagined for them by the pioneers of space travel.

Serge Brunier

Departure

NIGHT FLIGHT FOR THE AMERICAN SPACE SHUTTLE. LITERALLY TORN FROM THE EARTH'S PULL BY THE TREMENDOUS POWER OF ITS ENGINES, BUILDING UP TO A HUNDRED MILLION HORSEPOWER, *DISCOVERY* CARRIES ITS CREW OF SIX ASTRONAUTS TOWARDS THE MIR SPACE STATION. IN UNDER TEN MINUTES, *DISCOVERY* WILL LEAVE THE ATMOSPHERE AT A SPEED OF OVER 28 000 KM/HR AND GO INTO ORBIT AROUND THE EARTH.

■ ROBERT PARKER AND JOHN LOUNGE PREPARE FOR LIFT-OFF ABOARD
SPACE SHUTTLE *COLUMBIA*. IN A FEW MINUTES, THEY WILL CLOSE THE VISORS
ON THEIR HELMETS AND COUNTDOWN WILL BEGIN.

All things considered, it was perhaps for him that these moments were the most difficult. For them, fastened at the top for several hours now, jammed into their pressurised space suits, uncomfortably inclined and strapped against their seats, the countdown had begun only too long ago. They knew that the only way out was upwards. John Young, the most senior astronaut still in service and head of NASA's astronauts for over twelve years, was now director of the Johnson Space Center in Houston, Texas. At sixty-eight, and despite a staggering list of achievements as an astronaut, he could not help envying these men and this woman, twenty or thirty years his junior, about to open up a new era in the conquest of space. Right then, in the drab sunlight of an early winter morning at Cape Canaveral in Florida, John Young turned to face the huge digital clock as it slowly counted the seconds away. Behind him were the stands where VIPs had gathered since the legendary Apollo flights. In front, in line with the clock, was the space shuttle, upright on its launch pad, some 4500 metres away. Young was responsible for the six astronauts who were about to leave. As the minutes slipped away and the pressure mounted within, his face remained fixed in the confident smile of a commander for the benefit of officials and journalists, and especially for friends and families. Perched fifty metres up, the astronauts were as tense as the metal ribs of the shuttle's giant propellant tank, filled to bursting with liquid hydrogen and oxygen. They were not afraid. Or rather, they were afraid, that lift-off would be postponed once more. They would then have to wait, return to training and go through the motions already repeated a thousand times before the computer screen or in a swimming pool filled with the ghostly silhouettes of divers. And then they would have to begin everything again: the early morning rise, meticulous preparation, ritual journey by bus in the clear tropical night, and farewells just beside the launch vehicle as it towered impressively in the glow of the floodlights, an imposing and unlikely space vessel. But on 4 December 1998 the clock strung together the seconds without qualm or hesitation and did not stop minutes before blast-off because great storm clouds threatened the Everglades and Keys of Florida. And nor did it falter because one of the four onboard computers, in a hundred thousand exchanges per second with its three colleagues, had contested some instruction or some ambiguous result of a routine control. So, because all the

■ THE EARLY HOURS AT CAPE
CANAVERAL, FLORIDA. THE SHUTTLE
COLUMBIA IS READY FOR LIFT-OFF. THE
INAUGURAL FLIGHT OF THE VERY FIRST
AMERICAN SPACE SHUTTLE WAS MADE IN
1981, WITH JOHN YOUNG IN
COMMAND. IN 2000, *COLUMBIA* HAD
ALREADY ACCOMPLISHED TWENTY-SIX
RETURN TRIPS INTO SPACE. AFTER LESS
THAN ONE DECADE, ITS SOLID
ROCKET BOOSTERS, FILLED WITH
DANGEROUS POWDERS, WILL BE
REPLACED BY SAFER AND MORE
POWERFUL LIQUID HYDROGEN
AND OXYGEN BOOSTERS.

■ *ATLANTIS* ON THE
CRAWLERWAY AT CAPE
CANAVERAL, BEFORE JOINING THE
FAST LANE TO THE STARS. NASA
CURRENTLY MAINTAINS A FLEET
OF FOUR SPACE SHUTTLES. IN
THE YEARS TO COME, *ATLANTIS,
COLUMBIA, DISCOVERY* AND
ENDEAVOUR WILL TRANSPORT
ASTRONAUTS AND
EQUIPMENT UP TO THE
INTERNATIONAL SPACE
STATION.

■ CAPE CANAVERAL ENGINEERS ARE WITNESS TO A SCENE WORTHY OF A SCIENCE-FICTION FILM. TWO SHUTTLES, *COLUMBIA* ON THE LEFT AND *ATLANTIS* ON THE RIGHT, CROSS PATHS AS THEY ARE ESCORTED TO THEIR RESPECTIVE LAUNCH PADS.

indicators were green and nothing else might now prevent the launch, the space shuttle would soon wrench itself away from the surface of our planet.

THE FAST LANE TO THE STARS

It was just an ordinary flight. The ninety-third launch of an American space shuttle. However, it was not just a flight like any other. For one thing *Endeavour* was carrying Node 1, a large chunk of the future International Space Station, and so would contribute to the largest space project to be realised at the beginning of the twenty-first century. For another, a manned vehicle is not sent into space with the same ease or routine as a long-distance passenger flight destined for another continent. Technological precision, safety measures, weight and energy are all marked up by an order of magnitude. The shuttle has a lift-off mass of 2000 tonnes, on a par with eight huge Airbus A340s. The five engines about to spring to life would develop a thrust of around 3500 tonnes, equivalent to that of sixty Airbuses. Finally, when the outward flight came to an end 400 km above the Earth, the shuttle would have exceeded the unimaginable speed of 28 000 km/hr, thirty times that of an ordinary airline flight.

Moreover, such flights as this remain unique for quite another reason: the space shuttle has all the features of a decoy, with the streamlined profile of a glider and the illusory status of a space plane. Designers and users alike admit the dangers concealed by the marvellous machine. Over-complex and fragile, it calls for permanent attention, with long, delicate and costly inspections after each flight. Although designed to carry out return trips to space at lower cost, as its name would suggest, the ultimate paradox is that the shuttle has ended up being far more expensive than the rustic but reliable Russian Semiorka rockets. These are used only once and mass-produced, like Europe's Ariane launch vehicle. Could it be that the American space shuttle was born before its day? On 12 April 1981 when *Columbia* left Earth for the first time, the strange impression of this great white bird climbing vertically before gently leaning towards the east left a genuine foretaste of the twenty-first century. But alas! Expectations were soon brought back to Earth. Less than five years after NASA set up its science-fiction space transport system, with four vehicles *Columbia*, *Discovery*, *Atlantis* and *Challenger*, the *Challenger* shuttle exploded at lift-off and seven astronauts were lost. Today there are still four shuttles. Numbers have been made up by a new arrival, *Endeavour*. The fleet carries out its missions without a hitch but at a high price: something like 350 million euros for a one week flight, with a further 50 million euros for the three-month rest period between launches.

■ *COLUMBIA* ON ITS LAUNCH PAD A FEW HOURS PRIOR TO LAUNCH ON 22 JULY 1999. STILL IN PLACE, THE FOOTBRIDGE IS USED BY THE SEVEN ASTRONAUTS TO TAKE THEIR POSITIONS IN THE COCKPIT. THE X-RAY TELESCOPE CHANDRA IS CAREFULLY STOWED IN THE CARGO BAY OF THE SHUTTLE.

JOHN YOUNG AT THE COMMAND OF *COLUMBIA*

From afar, John Young watched over the launch pad where *Endeavour* stood so imposingly, immobile, ready for blast-off. To his mind, it is the night before the flight when the shuttle is most impressive, when the way to launch pad 39A is opened to guests, just prior to filling the main propellant tank. Powerful floodlights then light up the machine, bringing out its sleek profile and blurring the gigantic metal framework that still binds it to Earth. When they find themselves standing at the foot of this great white bird stranded in the night, spectators are struck by the beauty and solemnity of the sight. Without fail, they ask whether they themselves would be able to climb into the tiny space up there at the top. Just the thought of it sends a shiver of pleasure down the spine. But more often than not, they feel their legs giving way.

John Young could not remember how many shuttles he had watched in this way as they prepared to leave for space. However, he did recall with perfect clarity that spring morning when he himself was up there, suspended between heaven and Earth, waiting impatiently for lift-off. It was 12 April 1981, exactly twenty years after the flight of Yuri Gagarin, that John Young became captain of the very first space shuttle. The man is a kind of extraterrestrial with a quite unique experience. A symbolic figure in the conquest of space, his career will probably remain one of the finest human achievements of the second millennium.

It was in the spring of 1965, at the age of thirty-five, that this American astronaut first went into space, aboard the Gemini 3 capsule. The following year, during summer, Young went back into Earth orbit for the Gemini 10 mission. But his incredible adventure was only just beginning. For his third flight, in May 1969, John Young set off for the Moon with Tom Stafford and Gene Cernan, a space odyssey of more than a million kilometres, lasting a week. Aboard the Apollo 10 command module, the astronauts went into orbit around the Moon. For the crew, the mission was both wonderful and frustrating. John Young, Tom Stafford and Gene Cernan went through all the lunar landing procedures, even approaching to within 15 km of the surface of the Moon. Three years later, Young set off for the Moon yet again, this time in command. Having landed the Apollo 16 module in the region of the Descartes crater, John Young and Charlie Duke spent almost three days exploring the surface of another world with an electrically powered jeep.

With six flights, or seven counting the lift-off from the Moon, John Young is the man who has most often travelled in space. He twice experienced the weightlessness of Earth orbit, then fulfilled Jules Verne's dream by orbiting the Moon, before walking on its surface like Tintin. Two flights aboard the first space plane conclude this unbelievable list, making him a worthy forerunner of Flash Gordon and Luke Skywalker. It has been a quite astonishing and extraordinary career, more impressive and fruitful than those of Yuri Gagarin, John Glenn and Neil Armstrong put together, and yet the general public knows almost nothing of John Young. Such are the vagaries of posterity.

The last time John Young left for space, aboard *Columbia*, was in November 1983. He is still a member of NASA's astronaut corps and may well pilot other space shuttles in the years to come. But for the moment, on 4 December 1998, John Young could only watch, in close communion with his six young colleagues. Up there in the cockpit at the top of the futuristic *Endeavour*, Americans Bob Cabana, Rick Sturckow, Jerry Ross, Nancy Currie, Jim Newman and their Russian guest Sergei Krikalyov gazed out through the wide glass screen towards the zenith and the royal blue sky of Florida.

This was the first space flight for Rick Sturckow. Reclining on his seat, awe-inspired, the young astronaut of thirty-seven years finally realised that this time they were no longer going through the motions of a simulation. Indeed, beneath him, the shuttle seemed to come alive and groan as a muffled vibration shook it from top to bottom. *Endeavour's* main propellant tank, like a gigantic eggshell stuck onto the shuttle, was filled to the brim with 750 tonnes of liquid hydrogen

■ AT BLAST-OFF, THE THRUST OF THE SHUTTLE'S FIVE ENGINES EXCEEDS 3000 TONNES, EQUIVALENT TO THE POWER OUTPUT OF SIXTY JUMBO JETS. IN LESS THAN ONE MINUTE, IT WILL REACH AN ALTITUDE OF 10 000 M AND BREAK THE SOUND BARRIER.

and oxygen. The huge mass of its contents tended to distort its thin aluminium walls. From time to time, as though sighing, the great machine evacuated a blast of oxygen from the top of the tank via a relief valve. Half an hour before the engines were switched on, the launch pad was deserted. There was no one to be seen for three kilometres in any direction. Little by little, almost imperceptibly, the shuttle was already moving away from the flight control centre in Houston and the world of men, well before the lift-off itself actually took place. Slowly, the propellant feed mast and footbridge were withdrawn. Only five minutes remained. The shuttle now relinquished its electricity supply from the Cape Canaveral base. It would subsequently run on its own battery power source. The computers methodically continued their checking routines. The six astronauts in the cockpit could feel *Endeavour's* three main engines swivelling into the exact direction required for lift-off. Gradually, less and less information was exchanged between the Houston computers and those on board *Endeavour*. The time was 9h34 GMT. As launch approached, the shuttle had become completely autonomous and was almost ready to take control of its own destiny. On the giant clock at Cape Canaveral, John Young followed the countdown. Only thirty seconds to go. Now events began to move fast. Ten seconds. Although invisible to the spectators, hundreds of cubic metres of water were pouring down the concrete vents of the launch pad, submerging them in a torrent that would soon absorb part of the shock wave and the immense heat released by the shuttle engines. Without these measures, the heat and echo would rebound from the ground and might damage the shuttle during blast-off. Five seconds. The three engines were switched on. High-pressure liquid hydrogen and oxygen flooded into the combustion chamber and ignited.

ENDEAVOUR LEAVES EARTH AT 28 000 KM/HR

From this point on about 5000 litres of liquid propellant would be consumed by the shuttle every second. And yet the enormous thrust of the engines was still not sufficient to lift it from the ground. In the cockpit, the astronauts could hear the furious rumbling of the engines. They could feel their shuttle vibrating and oscillating slightly. Even at this stage, the onboard computers could still decide to close down the whole operation, by cutting the supply of hydrogen and oxygen and stopping the engines.

On the other hand, exactly two seconds beyond this point, the shuttle would be condemned to lift off. Indeed, it was the two solid rocket boosters flanking the main propellant tank that would provide the major part of the blast-off energy. Measuring 50 metres tall and weighing 590 tonnes, each booster is like a gigantic stick of solid propellant. Once lit, nothing can stop them. On this 4 December, however, the computers checked one last time that the shuttle's main propulsion system was functioning correctly and ordered the firing of the boosters. With its propellant tank full and its two boosters, *Endeavour* weighs slightly over 2000 tonnes. More than 1700 tonnes of this is propellant. A thrust of over 3000 tonnes is developed by the three engines of the main propulsion system and the two boosters together, as they spew out gases heated to 3200°C.

Once it had broken away from the moorings that held it to the launch pad, the great space bird could at last escape from Earth's gravitational pull.

Observed from the official stands, blast-off is thrilling, dazzling. In total silence, John Young and several hundred guests and journalists witnessed first a blinding flash at the base of the stationary shuttle, followed by disquieting swirls of smoke and water vapour, piling up like storm clouds around the launch pad. Silence still reigned as they observed the shuttle rising gently from this raging turmoil. It was only when *Endeavour* had climbed to several hundred metres altitude that the fearful roar of the hundred million horsepower unleashed during blast-off suddenly broke upon them, submerging all.

In the cockpit at this precise moment, Rick Sturckow was congratulating himself on the hundreds of hours spent practising this lift-off in a simulator. If this had not been the case, impressions would have been too strong and above all too unexpected. To begin with, there was a deafening noise accompanied by vibrations that seemed to shake the shuttle to the core as it tore itself away from the Earth. Then a giant's hand pushed, and pushed again, with a truly prodigious force. Rick Sturckow was crushed on his seat. Under an acceleration of 3 g (three times the acceleration due to gravity), he weighed over 200 kg. But worse, since the shuttle had automatically curved round and was now heading east towards the Atlantic, the astronaut was upside-down. And this was not the end of the ordeal. Less than one minute after lift-off, when *Endeavour* had already gone up over 10 000 metres, the engines suddenly slowed down and the spacecraft seemed to stop in full flight! The effect might have been singularly unpleasant, but Sturckow knew that all was going according to plan. In this phase of the flight, the shuttle was accelerating through a still dense atmosphere and undergoing tremendous mechanical stresses. At the critical moment when the sound barrier was broken, at 1100 km/hr, the computer cut the power output of the engines by a factor of almost two. Still thrust forward by the two boosters, now practically empty, the shuttle thus continued to accelerate but at a lower rate. At an altitude of 50 kilometres, the boosters were delicately jettisoned and the shuttle was now climbing at almost 5000 km/hr. Through the portholes, Rick Sturckow could see nothing. The sky had switched from blue to black. A second time, just as the boosters burnt out, he experienced the unpleasant but fleeting impression that something might be wrong, that the shuttle was going to stop. Then, once again, acceleration resumed, and this time the shuttle seemed to have brought off a great victory in its battle against the Earth's grasp. Flying well above the clouds in a considerably rarefied atmosphere, freed of its two boosters, it continued to accelerate gently, like an ordinary aircraft. But what an aircraft! Lift-off had occurred only six minutes previously, and yet *Endeavour*, propelled to almost 18 000 km/hr, was now at an altitude of more than 100 kilometres, somewhere over the Atlantic Ocean. In two minutes, the main propulsion system would be switched off. The large propellant tank, now empty, would be detached and jettisoned in its turn, at which point *Endeavour*, flying at over 28 000 km/hr, would leave the Earth's atmosphere and go into orbit.

■ AMERICAN ASTRONAUT STEPHEN OSWALD HAS FLOWN ABOARD THE SPACE SHUTTLE THREE TIMES. HERE THE *DISCOVERY* PILOT HAS JUST SUCCESSFULLY NEGOTIATED THE MOST CRITICAL PART OF ANY SHUTTLE MISSION: INSERTION INTO EARTH ORBIT.

At the same moment, several thousand kilometres away in the little town of Cocoa Beach on the Atlantic coast and at nearby Cape Canaveral, a stunned crowd stood face upwards, still gazing awe-struck at the enormous plume of white smoke that traced out a flimsy stairway between Earth and heaven.

THE MOSCOW-TASHKENT-ALMATY LINE

The railway line reached across the monotonous steppe of Kazakhstan. A few strands of grass emerged from between the sleepers, yellowed by the late autumn sunshine. In the cool of the afternoon, an ageless diesel locomotive stood motionless with its short chain of tanker wagons. Beyond lay the never-ending line from Moscow to Almaty via Tashkent. Not far away from the stationary goods train, another locomotive

■ Since Yuri Gagarin's flight in 1961, the faithful Semiorka rocket has taken over a hundred cosmonauts of all nationalities into space without mishap. All manned flights have left from the legendary Baikonur cosmodrome, lost in the Kazakhstan steppe.

was moving slowly forward along a neighbouring line, cautiously towing a single wagon. It would have been easy to follow this unlikely convoy at walking pace. Indeed, it was preceded by a solemn procession of spectators, technicians, photographers and a few soldiers. Now and again someone leant across the rail to place a coin worth a few kopeks, then stood back to wait until the passing convoy had crushed it wafer thin, before picking it up again, an instant good luck charm. The whole spectacle was quite bewildering, like some anachronistic scene painted by Enki Bilal. Behind the familiar outline of the locomotive lay a gigantic and incongruous launch vehicle, the faithful old Semiorka. Carefully poised on its long platform, it seemed to sleep. In fact, its tanks were empty and the great metallic frame, though 50 metres long, weighed only 32 tonnes. The first journey of the Semiorka rocket towards the stars lasted a full hour, the time required to take it three kilometres from there, to its launch pad at the Russian base of Baikonur. Once the showpiece of the Soviet empire, Baikonur is now located in the Republic of Kazakhstan and rented out by the Russians for 115 million euros a year. Baikonur, a genuinely holy place, was revered by an entire nation. It is a legendary name that recalls the glory of an empire now fallen into decline, but which was the first to reach the starry skies. However, it is also a secret place. The name is a bluff. The most secret military base in the Soviet Union, built in 1955 and situated at 46°25'N and 63°25'E, is actually close to the village of Tyuratam, more than 350 kilometres south west of Baikonur. Needless to say, this deceit, thought up by Red Army officials, did not long fool the sharp eye of US spy planes like the U2. At this historic site, in so many ways unchanged today, the conquest of space made its first decisive steps half a century ago. It was at the height of the Cold War, when two deeply opposed ideologies each sought supremacy, a battle of propaganda raged on both sides and a genuine technological war was engaged with a major military objective: control of space. On the American side, engineer Wernher von Braun was retrieved when the Germans surrendered, and sent to New Mexico to develop intercontinental missiles using his

experience of the V2 rockets built at the secret base of Peenemünde during World War II. On the Soviet side, the brilliant engineer Sergei Korolev also developed strategic missiles using recovered German V2 bombs.

The world-dominating ambitions of the two most powerful nations caused them to seek new horizons, and what could better symbolise their dream of greatness than the infinite horizons of space? In this exorbitantly expensive and planet-wide trial of strength, which the frail Soviet economy finally proved unable to sustain, it was nevertheless the USSR that won the first few rounds. At 19h 28m 34s GMT on 4 October 1957, the space age finally began. In Baikonur that autumn evening, Sergei Korolev gave the order to fire the very first Semiorka rocket. Weighing 270 tonnes at blast-off, it carried with it the first artificial satellite, *Sputnik*, a small sphere 58 centimetres across and weighing 83 kg. Inside were two small radio transmitters. This was the first time in human history that man could escape from the Earth's attraction, if only by proxy. From its highly elliptical orbit, taking it as far as 940 kilometres from Earth, Sputnik transmitted its historic beep-beep across a stupefied planet and a prostrate United States. Struck dumb by this extraordinary demonstration of know-how, the Americans had not yet seen the end of the lesson. On 12 April 1961, the very same Semiorka rocket carried the first astronaut into space. Yuri Gagarin became a genuine idol for the Russian people and, with hindsight, it must be said that the courage and confidence of the young cosmonaut, only twenty-seven years old, deserve admiration. Korolev's rocket into which he climbed that April morning had suffered several failures during lift-off tests and the first human space flight looked more like a round of Russian roulette. Statistically speaking, the Soviet fighter pilot had slightly less than one chance in two of returning alive from his tour of the stars. Yet Yuri Gagarin displayed an Olympian calm as he took his place in the capsule and the Semiorka blasted off. Throughout his automatically controlled flight, which ran without any particular problems, he noted down his impressions in a school notebook,

■ DESIGNED OVER HALF A CENTURY AGO BY SERGEI KOROLEV, THE SEMIORKA ROCKET HAS CARRIED OUT OVER A THOUSAND FLIGHTS TO DATE. EACH LAUNCH TAKES PLACE IN A TRULY CEREMONIAL ATMOSPHERE, A CONVIVIAL CELEBRATION IN HONOUR OF THE COSMONAUTS THAT LEAVE FOR THE STARS.

■ SURREALISTIC SIGHT IN THE DESERT-LIKE STEPPE OF KAZAKHSTAN AS A RAIL CONVOY SLOWLY GUIDES A SEMIORKA ROCKET TO THE LAUNCH ZONE AT BAIKONUR. AT THIS STAGE THE ROCKET IS MERELY A VAST EMPTY SHELL, 50 M LONG BUT WEIGHING NO MORE THAN 32 TONNES. WHEN IT STANDS UPRIGHT ON THE LAUNCH PAD AND ITS TANKS HAVE BEEN FILLED, IT WILL WEIGH 320 TONNES.

PARADOXICALLY, IT IS ONLY WHEN THIS HUGE MASS HAS BEEN REACHED THAT IT CAN ESCAPE FROM EARTH'S GRAVITATIONAL PULL.

only stopping when his pencil drifted across the weightless cabin and disappeared behind his seat. After a flight lasting 1h 48m during which he attained the extraordinary altitude of 327 kilometres and travelled more than 40 000 kilometres, Yuri Gagarin returned to Russia and landed on the banks of the Volga. Officially, it was the Vostok capsule, a huge metallic sphere, that brought him back to Earth, but in reality it is now known that Yuri Gagarin was ejected at an altitude of 7000 metres to land safe and sound on Russian territory with the aid of a parachute.

SEMIORKA, LEGACY OF SERGEI KOROLEV

The convoy had stopped. The railway ended on a huge expanse of concrete cut into the side of the hill. Technicians were passing to and fro in an almost routine fashion, watching calmly as the rocket slowly righted itself on the platform, lifted by a powerful mechanical leverage system. Once erected, four metal arms gradually closed in and delicately embraced

■ NOZZLES OF THE FIRST AND SECOND STAGE ENGINES OF THE SEMIORKA ROCKET. DURING BLAST-OFF, THE MOST RELIABLE ENGINES IN THE HISTORY OF SPACE FLIGHT UNLEASH A THRUST OF 470 TONNES.

the Semiorka, holding it securely in this position until blast-off. It was now a fine sight, dominating the Kazakhstan steppe. Around it, the largest space base in the world stretched away as far as the eye could see. Baikonur covers 10 000 square kilometres, with more than 1200 kilometres of road, 470 kilometres of railway line, a dozen or so civilian launch pads and military silos, and a town, Leninsk. The latter was built some 40 kilometres away to house the workforce from the base and their families, a total of around 50 000 people.

Once the rocket had been stood upright, two days before the launch, all that remained was to carry out a complete check of its flight systems, connections between its three stages, and the correct functioning of its five engines, which include no fewer than thirty-two separate combustion chambers. The Russian rocket is indeed quite different from its American rival. Although almost four times heavier, the shuttle flies with just five engines. In the 1950s, Korolev had not yet mastered the

24

■ VALERI KORZUN AND ALEKSANDR KALERI ARE READY FOR LIFT-OFF ABOARD THEIR SOYUZ. FOR AROUND TWENTY YEARS, THE RUSSIANS HAVE WORKED WITH COSMONAUTS FROM A DOZEN NATIONS.

sophisticated technology required for such technically demanding and powerful engines. In the circumstances, he preferred to increase the number of combustion chambers, making them smaller and clustering them together in groups. The Semiorka engines produce a thrust of around 25 tonnes, compared with 210 tonnes for the shuttle engines, currently the most powerful in the world. The fuel used by the Russians is also rather unsophisticated – a simple form of kerosene! On the other hand, Russian cosmonauts have every reason to trust their old launch vehicle. Indeed, at least a thousand flights have been successfully accomplished by successive generations of Semiorka rockets since 1957, and Korolev's successors can pride themselves on never having lost a single cosmonaut aboard this rocket. Only twice did they come close to disaster. In April 1975, the third stage of Soyuz 18A failed to separate from the second stage but still ignited. The mission was aborted at 192 km altitude and the capsule landed in the Altai mountains. A more serious incident occurred in September 1983. Cosmonauts Vladimir Titov and Gennadi Strekalov were waiting calmly for blast-off in their Soyuz capsule, but when lift-off finally took place, it was not at all as they had imagined it. A minute and a half before the engines were switched on, fire broke out in the first propulsion stage of their Semiorka. In an instant, the rocket was transformed into a giant incendiary bomb, threatening to explode at any moment. The cosmonauts triggered their ejection system, a string of explosives which flung them 1200 metres up into the air within a few seconds. Down below, all hell was let loose, as Semiorka's 290 tonnes of propellant went up in a gigantic explosion. The two men landed several kilometres away, alive but in a state of severe concussion. During the explosive ejection from Soyuz, they were briefly subjected to an acceleration of 15 g.

■ THE METAL ARMS THAT HOLD THE SEMIORKA SWING BACK LIKE THE PETALS OF A FLOWER TO RELEASE THE ROCKET. IT CAN NOW LEAVE EARTH, CARRYING VIKTOR AFANASYEV, IVAN BELLA AND JEAN-PIERRE HAIGNERÉ OUT TO THE MIR SPACE STATION.

ON ROUTE FOR THE INTERNATIONAL
SPACE STATION

Today cosmonauts Sergei Krikalyov and Yuri Gidzenko, together with guest American astronaut Bill Shepherd, were preparing to leave the Baikonur cosmodrome. Bill Shepherd recalled for a moment that the rocket standing before him was the very same Semiorka that had carried Yuri Gagarin into space forty years earlier. As the great machine seemed to pour out smoke from all quarters, continually venting out its excess liquid oxygen, the two Russians and the American could feel the same vibrations. They perfectly symbolised the only two nations currently capable of sending humans into space. Moreover, Sergei Krikalyov had known his American colleague for a good many years. He had trained with him at the Houston training centre and also at the Star City training centre near Moscow. Only forty-two years old, Krikalyov was already a veteran. He had flown twice aboard Mir, spending a total of more than a year in space. In addition, he had taken part in two shuttle flights, aboard *Discovery* in 1994, and aboard the famous *Endeavour* flight of 4 December 1998 which undertook the first journey out to the new International Space Station.

The rocket was now ready. All its tanks had been filled with kerosene and liquid oxygen and it weighed almost 320 tonnes. The three men at the foot of the launch vehicle climbed a few steps and entered the lift that would take them up to their Soyuz craft at the top of the Semiorka, two hours before launch. Although less spacious than the shuttle cockpit, its unsophisticated and robust appearance was reassuring. In addition, the ultracontinental climate of Kazakhstan starkly contrasted with the vicissitudes of weather conditions in Florida. Rockets leave Baikonur night and day, summer and winter alike. Come what may, a Semiorka will fly at any temperature between −35°C and +40°C.

Around the launch pad, spectators moved back at the request of debonair soldiers. Security is much more relaxed and hospitable than at Cape Canaveral, with its heavy police presence. Most of the journalists and a handful of courageous guests gathered about 500 metres from the launch vehicle, whilst a Russian photographer, ears filled with wax, literally risked his life by climbing into a small box in a communications tower less than 100 metres from the pad. As lift-off approached, the Semiorka was gradually gaining independence from its flight controllers, who had taken up positions in a concrete bunker. Thirty minutes to go. Slowly, the tall access and propellant feed towers moved away and delicately folded down to the horizontal. Seven minutes to go. All the timing of the launch was automatic from now on. Various electrical umbilicals, cables linking the rocket to its base, were now detached, thus freeing the Semiorka, apart from the four clamp arms that would maintain it in the upright position until the very last moment. Ten seconds. The engines of the four lateral boosters were fired. The monstrous energy released by burning gases expelled at 3 km/s was evacuated via a huge concrete flue inclined at 45° and dug straight into the hillside.

The engine of the central stage was then switched on and in a few seconds the total thrust of the thirty-two combustion chambers reached 470 tonnes, exceeding the rocket's own weight. Very slowly it began to rise. As it did so, it pushed aside the four clamp arms, and they swung open like the petals of a flower under the effects of a counterpoise system. Semiorka was free at last. With Sergei Krikalyov, Yuri Gidzenko and Bill Shepherd aboard, it was climbing at several hundred kilometres an hour. It would soon break the sound barrier and head away to the distant orbit where the beginnings of the International Space Station had been patiently awaiting it for the past two years.

■ ACCOMPANIED BY A CACOPHONOUS BLAST THAT FILLS THE AIR WITH A MUFFLED VIBRATION, SEMIORKA REACHES UP TOWARDS THE SKY. FOR A BRIEF INSTANT, THRUST UPWARDS BY A BLINDING FLAME, THE ROCKET SEEMS TO CHALLENGE THE BRILLIANCE OF THE SUN.

Barren void

THE BEAUTY OF THE WORLD AS SEEN BY ASTRONAUTS IN ORBIT. HERE, SWISS ASTRONOMER CLAUDE NICOLLIER HAS JUST FINISHED WORKING ON THE HUBBLE SPACE TELESCOPE AND TAKES THE OPPORTUNITY OF PHOTOGRAPHING THE COAST OF CHILE. CLOUDS GATHER ABOVE THE SOUTH PACIFIC BUT NEVER MANAGE TO REACH THE ATACAMA DESERT. BEYOND, A FEW CLOUD FORMATIONS CLING TO THE HIGH ANDES.

■ ASTRONAUT LOREN SHRIVER OBSERVES THE EFFECTS OF WEIGHTLESSNESS ABOARD THE SHUTTLE *ATLANTIS*. IN SPACE, THE CHOCOLATE DROPS FOLLOW A STRAIGHT LINE WHEN THROWN.

Sergei Avdeyev observed the perfectly immobile blob of coloured life. It was floating freely in space, smooth, round and green, strangely lost amidst the ill-assorted jumble of electronics. On one side a portable computer with electricity cables, strips of Velcro, and various tubes emerging from it, on the other, a bicycle leaning over at right angles, and family snaps fixed on a partition wall. Beyond, a forgotten notepad was gently spinning about its own axis. Dressed in shorts, tee shirt and thick woollen socks, crouching with knees pulled up to his chest as though in levitation, Sergei Avdeyev watched the solitary apple.

This was an almost priceless gift to someone who had been living here for over seven months, like Sergei Avdeyev. Aboard Mir the cosmonauts did not often have the opportunity to taste the wonderful fruits of their blue planet, that other life-containing orb that slipped past the space station windows. The apple had arrived on the Progress resupply craft along with a stock of fuel, food, water, clothes, spare parts for computer repairs, batteries, and mail. There were even small presents promised to the three cosmonauts by the Moscow technicians at the TSOUP, the ground control centre for the space station, located at Kaliningrad. Progress 41 had docked with Mir a few hours

previously. Following the tension of the docking manoeuvres, so often tricky or even harrowing, Sergei Avdeyev was now relaxing with his two companions, the Russian Viktor Afanasyev and Frenchman Jean-Pierre Haigneré. This was his third space flight. Selected in 1987, the young cosmonaut had had to wait six years before finally obtaining a place aboard the venerable Semiorka, in June 1992, to make his first flight out to Mir. He had spent more than six months here on that trip, including two weeks with French cosmonaut Michel Tognini. In 1995, he had returned to Mir for a further six months. By then he had fallen in love with his space dacha, and in August 1998, had set off for the stars once more. At the age of forty-three, he had become one of the most seasoned cosmonauts. Within a few months, when he returned to Earth on 28 August 1999, he would be the man who had spent longest in space, totalling 748 days, or more than two years in orbit, the record for the second millennium. In his company, Jean-Pierre Haigneré would also beat a much coveted record. The French cosmonaut, on his second flight aboard Mir, would become the first non-Russian cosmonaut to have stayed more than six months in space. This was an invaluable experience for the whole body of European 'spationauts', soon to move in to the International Space Station.

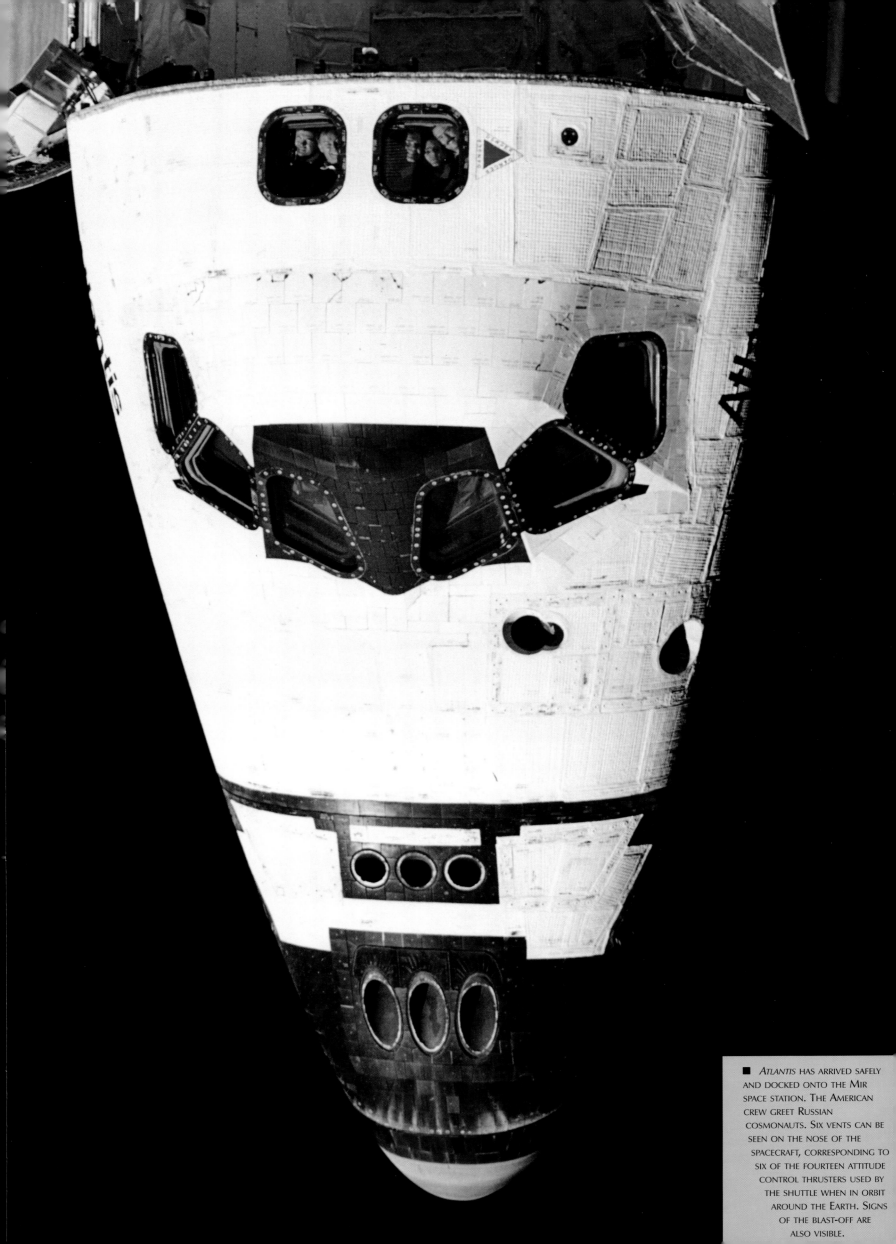

■ *ATLANTIS* HAS ARRIVED SAFELY
AND DOCKED ONTO THE MIR
SPACE STATION. THE AMERICAN
CREW GREET RUSSIAN
COSMONAUTS. SIX VENTS CAN BE
SEEN ON THE NOSE OF THE
SPACECRAFT, CORRESPONDING TO
SIX OF THE FOURTEEN ATTITUDE
CONTROL THRUSTERS USED BY
THE SHUTTLE WHEN IN ORBIT
AROUND THE EARTH. SIGNS
OF THE BLAST-OFF ARE
ALSO VISIBLE.

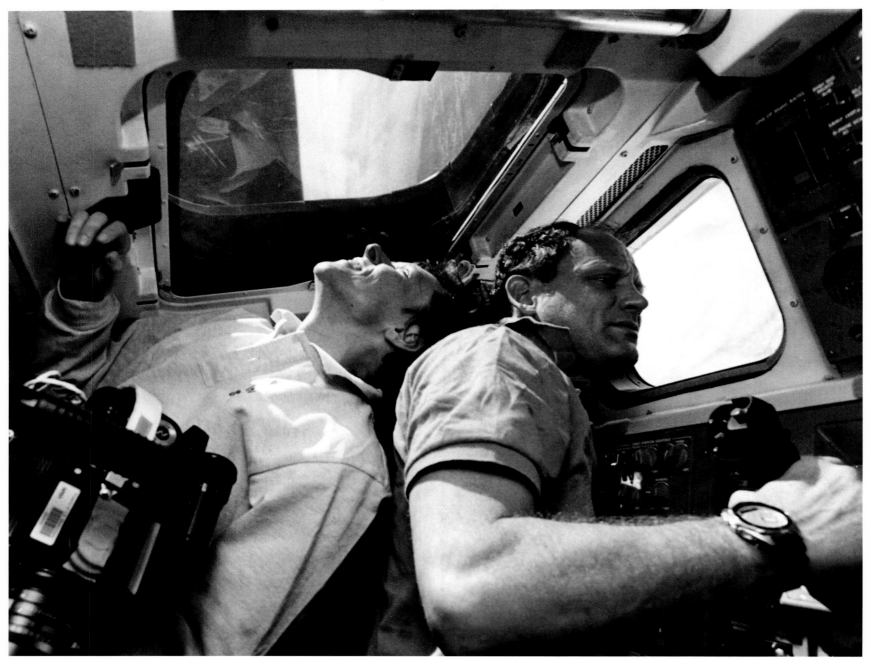

For the moment Sergei watched the apple, still perfectly motionless, floating serenely opposite him in the still air of the space station. But was it motionless? Through the porthole, the cosmonaut could see the Earth slipping past at a tremendous rate. Mir took only an hour and a half to complete its orbit, 350 kilometres above Earth's oceans and mountains. Motionless? Mir was moving at some 28 000 km/hr relative to our planet. The Earth is itself dragged at 30 km/s around the Sun, whilst even the Sun hastens towards the constellation of Hercules at 1 million km/hr through its motion around the galactic centre. The story does not end there either. The galaxy to which our star belongs, the Milky Way, is orbiting about its common centre of gravity with neighbouring galaxy Andromeda, and this pair, together with several dozen small satellite galaxies, is hurtling at 360 000 km/hr across the cosmos towards the Virgo cluster, some 50 million light years away. Finally, for the last fifteen billion years or so, the Virgo cluster has been retreating with ever increasing speed – several hundred km/s – from all other clusters of galaxies in the Universe.

It is a stunning, but unbelievable cosmic roundabout. What cosmonaut aboard his or her frail vessel has not reeled before this vast realisation? Where better than this, in the cold emptiness of space, could we penetrate the genuine mystery of our own existence, in this basically empty Universe? As Leibniz asked 300 years ago: Why is there something here, rather than just nothing? And why does it obey the laws of nature, so disconcerting, no more than intuitive, indeed incomprehensible, and yet directly tangible to the cosmonaut up here, floating in a space station?

The feeling of weightlessness, the absence of gravity, is indeed an unimaginable experience. It is quite impossible to communicate it to someone who remains trapped upon the Earth. No experience on our planet can help us to understand it. The evolution of human beings is deeply rooted in the vertebrate division, with which we have a long-standing affiliation, taking its source in the very beginnings of life, during the Precambrian and Palaeozoic, five hundred million years ago. Since then, the anatomy of the long succession of animals leading to ourselves – first fish, then amphibians and finally mammals – has evolved in response to this force, our weight, borne by our shoulders and holding us to the ground. It is an innate force, present in some latent form in every one of our cells, since the beginning of time. A vertical reference, subconsciously influencing our every act. To violate this law is

to trip and fall, to hurt ourselves or even die. Understanding weight, by imagining life without it, is a considerable intellectual undertaking. It is no surprise then that it has taken four centuries, if we consider the Renaissance as the origin of experimental science, to begin, just begin to comprehend what happens when cosmonaut Sergei Avdeyev floats beside a stationary apple in a spacecraft hurtling round the Earth at 28 000 km/hr.

An apple in the moonlight

In a small orchard not far from Cambridge, Isaac Newton observed the apple, motionless upon its branch. It was August 1684, at the close of afternoon. Suddenly, the apple broke from its branch and fell. To the south east, far beyond this famous apple tree, a beautiful full Moon was rising slowly into the sky. So it is said that the brilliant English mathematician and physicist discovered the law of gravity, in a flash of intuition that applied to the whole Universe. The apple and the Moon are subject to the same force. If the Moon forever turned about the Earth, thought Newton, it was because its natural motion exactly counterbalanced Earth's gravitational pull. More fundamentally,

the path of the falling fruit and the orbit of the Moon were of the same nature. So, if the apple were thrown with sufficient force, it would not fall back to Earth, but merely orbit around it, just like the Moon. Conversely, if the Moon did not have sufficient motion, it would fall onto our own planet, just like the apple.

In 1687, Newton published his *Philosophiae naturalis principia mathematica* (Mathematical Principles of Natural Philosophy, often referred to as Newton's *Principia*), a work which founded modern cosmology and at the same time pointed the way to a future space odyssey. Indeed, the universal law of gravitation set out by Newton, which says that all bodies attract one another with a force that is directly proportional to the product of their masses and inversely proportional to the square of the distance between them, really refers to all bodies in the Universe, whether they be galaxies, stars, planets, natural or artificial satellites, apples on board spacecraft or apples falling from trees. Moreover, Newton's equations are still used three centuries after the mathematician first formulated them. To cast an astronaut into orbit around the Earth or to fling a space probe far across the Solar System, Newton's law of gravitation is quite accurate enough although, as we shall see, the magnificently

ordered world of Newton concealed some grey areas and would finally give way to a new and even deeper vision of the Universe, more general than any before it.

Exactly 270 years after the publication of Newton's *Principia*, the mathematician's beautiful thought experiment could at last be turned into a reality. Like the apple that Newton had suggested hurling into the heavens with such great force, Sputnik was launched into space by a phenomenally powerful sling, the first Semiorka rocket. In order to tear itself from Earth's gravitational pull, any body, whatever its mass, must attain a speed of at least 28 000 km/hr, sufficient for it to go into orbit. On 4 October 1957, Semiorka finally achieved Newton's dream, carrying Sputnik well above the Earth's surface and into space.

But what is space? In Newton's day, the idea of a cosmic vacuum was just beginning to make its way into the accepted scientific culture. The invention of the barometer showed that atmospheric pressure decreases with altitude. It was gradually realised that somewhere above the highest mountain ranges, there might be a region dominated by empty space. Indeed, this is a necessary condition for the heavenly bodies to move freely around one another. If the Moon had to push the atmosphere aside on its path, it would lose energy and end up falling to Earth. In practical terms, it is the atmosphere which prevents us from inserting observational instruments into orbits lower than 100 kilometres altitude. Spy satellites, for example, are slowed down by this frictional effect and fall to Earth after a few days or weeks, unless equipped with an engine capable of periodically boosting them back into their nominal orbit. Even between 300 and 400 kilometres up, the Mir space station has suffered from such atmospheric braking since it was placed in orbit in 1985. A Progress spacecraft was regularly used as a kind of celestial tugboat to pull it back into orbit.

Isaac Newton's universe was infinite. Space is empty, providing a rigid and independent framework for the objects moving through it. It is a static stage on which history's theatrical performance is played out, from the rhythm of the celestial spheres to the fall of apples. To this absolute space there corresponds an absolute time, a kind of universal cosmic clock. Finally, for Newton, gravity was a force, propagating at infinite speed from one body to another. In addition, Newton's universe has to be boundless: this is the logical prerequisite for the universal theory of gravitation. As Newton noted, all bodies in the Universe attract all other bodies. Therefore, if the Universe were not infinite, homogenous and isotropic, these

■ A DROP OF WATER IN ORBIT IS SPHERICAL, LIKE A BLUE PLANET.

bodies would ineluctably fall toward one another, until finally they collided and formed a single object. Since this is clearly not what happens in the starry sky we observe, the Universe must be infinite. It is the very condition for its equilibrium.

NEWTON'S INFINITE AND EMPTY SPACE

The absolute space and time giving form to Newton's vision of the cosmos are an obvious and profoundly intuitive referent that is perfectly familiar to us. What more natural when we raise our eyes to the sky than to imagine ourselves in the middle of an infinite empty space, infinitely populated with stars? After all, is this not the view perceived by Sergei Avdeyev, Viktor Afanasyev and Jean-Pierre Haigneré when they looked through one of the portholes of their space station?

During the eighteenth and nineteenth centuries, Newton's law of gravity was put to use with tremendous success. Not only did it explain the motion of the heavenly bodies to very great accuracy, but it also proved to have extraordinary predictive power. The discovery of the planet Neptune by astronomers John Adams and Urbain Le Verrier in 1845 is a prime example. They independently announced its existence on the sole basis of gravitational perturbations observed in the orbit of Uranus. The new planet predicted by the calculation was observed by astronomer Johann Galle in Berlin shortly afterwards, using a refracting telescope. However, as time went by, things began to grow more complicated. Isaac Newton's sublime machinery began to go awry. To begin with, physicists discovered the wave nature of light and decreed that a wave required a medium in order to propagate. This spelt the end for empty space. For several decades, the idea of an ether was reintroduced. It was the fifth element, invisible and perfect, as conceived by Aristotle two thousand years before, to constitute the various nested spheres carrying the planets and stars. By the end of the nineteenth century, the ether had become an imponderable and mysterious medium permeating the whole Universe and physicists were keen to demonstrate its existence experimentally. The idea was to measure the Earth's motion relative to the putative medium. It appeared that nothing could be simpler. Albert Michelson and Edward Morley set up a method for accurately measuring and comparing the speed of light in two opposite directions aligned with the direction of the Earth's motion as it orbits the Sun: light from a point in space towards which the Earth is moving, and light from another point from which the Earth is moving away. The Earth orbits the Sun at 30 km/s whereas light travels at around 300 000 km/s.

■ Four cameras float almost motionless around Terence Wilcutt. As suggested by Albert Einstein at the beginning of the twentieth century, the astronaut could carry out the same experiment in a freely falling lift on Earth.

Michelson and Morley used an interferometer, an optical instrument capable of extraordinary precision. They expected to find a difference of 60 km/s in the two speeds. The experiment was carried out on several occasions from 1881 and systematically yielded the same incomprehensible, even unacceptable result. It was the only result that absolutely no one at all had expected. The result was negative. What the Michelson interferometer kept stubbornly telling the world was that light always travels at the same speed, no matter what the speed of the emitter and the receiver. In other words, light always travels at 299 792.458 km/s.

These measurements, so carefully made and so accurate, were long considered to be wrong merely because they so blatantly violated one of the most natural and intuitive laws of physics: the additivity of velocities. Michelson and Morley's contemporaries thought that when someone walks at 5 km/hr up the carriage of a steam train which is itself moving along at 60 km/hr, that person must be moving at 65 km/hr relative to the stationmaster waiting on the platform. Conversely, if the traveller were to walk down the carriage towards the back of the train, his or her speed would be 55 km/hr relative to the platform. But what would happen if the traveller were carrying a lantern and shone it towards the stationmaster? Equipped with a Michelson interferometer, the latter would see the train passing at 60 km/hr, but the light rays emitted from it would nevertheless arrive at the same speed, viz., 299 792.458 km/s. By the end of the nineteenth century, there was no escaping the fact that the speed of light is constant, whatever the conditions of emission and reception. At the beginning of the twentieth century, this observation, the invariance of the speed of light, which had triggered a major crisis in classical physics, led one man to completely overturn our view of the world.

GALILEO DISCOVERS THE PRINCIPLE OF INERTIA

Albert Einstein was to throw the ether out of the train window, along with the Newtonian concept of space and time. He was not alone, of course. In particular, French mathematician and physicist Henri Poincaré contributed in an essential way to the special theory of relativity, although he is somewhat unfairly neglected in this respect. In order to break away from classical mechanics, Einstein sought his new vision of the Universe in an idea that considerably predates Newton. The inventor of the profoundly abstract notion of relativity was the Florentine physicist Galileo Galilei (1564–1642), often considered as the

first modern scientist. Working at the beginning of the seventeenth century, Galileo did not have an orbiting space station at his disposal. However, his experimental investigations of motion, acceleration and forces revealed the great majority of phenomena experienced by today's cosmonauts. An example is the perfect immobility of an apple in the Mir space station. Galileo tried to formalise the laws of mechanics. In particular, since he claimed that the Earth spins about its own axis and simultaneously orbits around the Sun, he wished to understand why these extraordinarily rapid motions were in no way felt by its inhabitants. What he discovered was indeed revolutionary. In order to put across the basic premises of relativity to his contemporaries, he described an imaginary journey aboard a boat pushed by a light breeze across a sea of oil. As Galileo explained, a passager in the hold would not realise that he or she was moving. All sorts of physical experiments could be carried out but the passenger would never be able to detect the motion of the boat. If a cloud of insects were released, they would fly out in all directions. An apple thrown vertically upwards would drop back down into the passenger's hand, even if the boat were sailing along at five knots and might have covered several dozen metres whilst the apple was in movement. Everything would

happen as though the boat were not moving. 'Motion is nothing', concluded the Italian savant. It was a brilliant insight that founded modern physics and at the same time laid the first stone in the great edifice of relativity theory.

Thinking about the two fundamental states of rest and uniform motion in which a body might find itself on our planet, Galileo finally concluded that nothing could distinguish them. The laws of physics were strictly the same on a moving boat as on a boat at anchor. The state of uniform motion is natural in the Universe and requires no further physical explanation. This is the principle of inertia: a body will continue indefinitely in uniform rectilinear motion if unperturbed. The discovery has even greater merit when we consider that, in the Renaissance, Galileo had to take air resistance into account. Because of friction with the air, it was actually impossible to demonstrate the principle of inertia directly! Today, of course, we are not the least bit surprised to observe that a space probe launched out to the limits of the Solar System will continue forever along the same path. Since this time, physicists have used the terms 'Galilean frame' and 'inertial frame' to describe reference systems in uniform motion, whatever the velocity with which they may move relative to one another.

This view of things proved perfectly satisfactory until the advent of the Michelson–Morley experiment, with its demonstration that the speed of light is invariable. It was Albert Einstein who rescued twentieth-century physics from the crisis, after years of reflection, when he finally managed to cut away the web of prejudices man has built up since the beginning of time. He extended Galileo's concept of relativity (movement is nothing) to space and time. In the special theory of relativity published in 1905, Einstein did away with the absolute framework imagined by Newton, that somehow lay outside physics, and replaced it by variables that were intrinsically physical. Space and time are relative to the velocity of the observer. The length of an object changes with its velocity, whilst the duration of a time interval is different for the person who experiences it and another observer moving past with a different velocity.

Although the ideas required to understand special relativity are counter-intuitive and complex, it is curious to note that the celebrated Lorentz transformations that govern the mathematics are relatively easy to apply. To simplify, we may say that in the physics of special relativity, when speed increases, space contracts and the flow of time slows down. This happens in such a way that the speed of light, which is the ratio of a spatial interval to a time interval, remains the same. The change of paradigm Einstein put forward is quite extraordinary, and yet within a few years the scientific community had largely adopted it. This was simply because all its theoretical predictions were confirmed one after the other with an almost supernatural level of agreement, both by laboratory tests and

■ In the vacuum of space, the aerodynamic shape of the space shuttle serves no purpose. Perched at the end of Challenger's remote manipulator arm, Bruce McCandless inspects the underbody of the shuttle with its two cargo bays wide open.

observations of the great natural laboratory provided by the Universe itself. Special relativity thus imposed a radical change in the framework of physical thought. Not only did space and time become physical variables, but in addition, they became indissociable. Albert Einstein's universe is a four-dimensional space–time.

The speed of light, which is also the speed of propagation of gravitational effects, and more profoundly, the maximum speed at which any information can be transmitted, plays a pivotal role, in such a way that the laws of physics are the same everywhere. It should be noted that the numerical value of this speed, 299 792.458 km/s in the vacuum, is quite arbitrary. It merely results from the units used historically to measure it. In fact, according to the theory of relativity, it is a limiting speed that cannot be

■ BRUCE MCCANDLESS HURTLES AROUND THE EARTH AT 28 000 KM/HR IN THE WAKE OF THE SHUTTLE, BUT HE HAS THE IMPRESSION OF BEING PERFECTLY MOTIONLESS. PERHAPS HE IS RECALLING GALILEO'S MARVELLOUS INTUITION BACK IN THE SIXTEENTH CENTURY: MOTION IS NOTHING.

exceeded by any massless particle, such as the photon, and that cannot be attained by any massive particle, such as the atoms making up a rocket and its occupants. From a physical point of view, the speed of light is rather like an infinite speed. In a way, it is the last remaining absolute value that Einstein handed down to physics after his great clean-up. At the end of this book, we shall return to the consequences of the relativistic equations for future interstellar space travellers.

Despite its many successes, the special theory of relativity only applied to Galilean frames, that is, systems in uniform motion relative to one another, and did not answer a deep question asked by Einstein, and indeed by all physicists since Galileo himself. Not content to rest with the discovery of the first relativity principle, Galileo had noted a fascinating and mysterious property of nature. In a now famous

■ SO FAR AWAY AND YET SO NEAR. ASTRONAUTS OBSERVE THE EARTH WITH BINOCULARS, DISCOVERING AN EXTRAORDINARY WEALTH OF DETAIL. AS THOUGH THEY WERE FLYING OVER THE SURFACE OF OUR WORLD IN A PLANE, THEY CAN MAKE OUT THE VARIOUS STREETS AND DISTRICTS OF THE CITY. THIS IS MIAMI ON THE FLORIDA COASTLINE, PHOTOGRAPHED FROM THE SHUTTLE *COLUMBIA*.

■ EARLY MORNING ABOVE THE SOUTH
PACIFIC, JUST OFF THE COAST OF NEW
ZEALAND. WITH ALL ITS SOLAR SAILS
DEPLOYED, THE WONDERFUL RUSSIAN
SPACE CARAVEL WILL SOON DOCK
WITH *ATLANTIS*. ON BOARD THE
SHUTTLE, SHANNON LUCID ADMIRES
HER FUTURE HOME IN SPACE. THE
ASTRONAUT WILL REMAIN ON MIR
FOR SIX MONTHS, A NEW RECORD
FOR AN AMERICAN ASTRONAUT.

■ IN THOUGHTFUL MOOD, COSMONAUT VALERI KORZUN VENTURES INTO THE VERY HEART OF THE RUSSIAN SPACE STATION. THE CONNECTION MODULE HAS SIX AIRLOCKS AT WHICH THE *PRIRODA*, *KVANT 2*, *SPEKTR* AND *KRISTALL* MODULES ARE DOCKED, TOGETHER WITH A *SOYUZ* SPACECRAFT. VALERI KORZUN IS LEANING ON THE AIRTIGHT TRAPDOOR OF AN OPEN AIRLOCK. CABLES AND VENTILATION TUBES WEND THEIR WAY IN ALL DIRECTIONS.

but probably imaginary experiment, he suggested dropping various objects from the top of the Leaning Tower of Pisa, and watching them carefully as they fell. It turns out that, when the bodies are such that air resistance can be neglected, they all hit the ground at exactly the same moment. We are so used to observing this phenomenon in everyday life that we simply take it for granted. It seems perfectly natural to us. But let us follow Galileo to the foot of the leaning tower: the 200-gram green apple and the 10-kilogram wooden ball move along the same trajectory in the same time. Whatever a body's mass, whatever its composition, gravity treats it the same way. Why?

THE ASTRONAUTS IN ALBERT'S LIFT

This question hung in the air for three centuries, before Einstein finally answered it with one of the most fertile intellectual visions of all time: the general theory of relativity. It was while thinking over Galileo's experiment that Einstein saw the way to generalising his special theory of relativity to the whole Universe. Like Newton's theory, the general theory of relativity is basically a theory of gravity. Einstein did not imagine himself at the foot of the Leaning Tower of Pisa. Instead he locked a physicist in a stationary lift at one of the upper floors of an enormous skyscraper. For the sake of his idealised mechanics experiment, he removed the atmosphere along with its perturbing and untimely frictional effects. Inside the lift, the experimenter observes an apple, without knowing where he or she is located. On the floor is a notepad. Einstein presses the button to send the lift to the top floor and it begins to rise. The physicist, completely enclosed within, does not feel any upward movement, but instead measures an increase in his or her weight. The weight of the apple in the experimenter's hand also increases. What is actually a movement of the lift is thus interpreted by the physicist as an increase in the force of gravity! A mischievous assistant then cuts the lift cable and it begins to fall. Inside the lift, the experimenter, still stationary in his or her frame of reference, is amazed to observe the apple and notepad floating about unsupported. Indeed, the experimenter no longer feels his or her own weight. When the lift is freely falling, it is as though

■ COMFORTABLY INSTALLED BEHIND THE CABIN WINDOW OF *ATLANTIS*, TERENCE WILCUTT ADMIRES THE RUSSIAN SPACE STATION. THE RUSSIAN–AMERICAN COOPERATION ON BOARD MIR HAS BEEN ALL-IMPORTANT IN MAKING TECHNICAL CHOICES FOR THE INTERNATIONAL SPACE STATION.

weight and hence gravity both disappear. What the physicist represents in his or her closed environment is precisely what Mir cosmonauts experience all the time. So true is this observation that today physicists use very tall evacuated towers to produce, for a few seconds, a low-cost version of the microgravity conditions prevailing in space. When Albert Einstein published his general theory of relativity in 1915, ten years after the special theory, not only were space and time relative to local conditions, but so was gravity itself. Going further, he declared that gravity was not a force at all, but rather a geometric feature of space–time, namely, its curvature. This curvature varies with the mass contained within the space–time. It is zero in an empty space–time and maximal in the vicinity of more massive bodies, such as stars. According to general relativity, we may say today that the mass of the Earth (6000 billion billion tonnes) locally curves the region of space–time in which it is located. This curvature resolves the mystery that fascinated physicists since Galileo's experiment. All bodies whatever their mass and chemical composition are treated equally by gravity for the simple reason that gravity is not a force but rather a field, of purely geometric nature. Hence, whenever we feel the pull of a weight on our shoulders, we may imagine (and it is not easy) that we are living in a world that is somehow leaning over, and that we must continually struggle against the slope. Likewise, we must abandon the idea of a force holding the Moon and Earth in mutual orbit around one another. The Moon is not attracted by the Earth as might be claimed in a Galilean system. Rather it is moving in a straight line, but in the curved space–time caused by the presence of planet Earth. The same can naturally be said for cosmonauts living aboard Mir. They too are following a geodesic, a shortest path in the curved space–time around Earth, the next best thing to a straight line when the geometry is flat.

It may be asked what could be the applications for such an abstract theory, so elegant and yet so much less intuitive and more complex than the classical theory of Newton. In fact, for all the everyday situations we may encounter on a human or even a terrestrial scale, Newton's theory proves to be perfectly adequate. Among other things, it is sufficient for sending rockets, manned space stations and unmanned probes into space.

However, it ceases to be valid when physical conditions reach certain extreme levels. When the gravitational field becomes intense, when speeds approach the speed of light, so-called relativistic effects come into play, and then only the general theory of relativity is capable of explanation and prediction. As an example, it makes a tiny correction to Newton's theory in the description of Mercury's orbit through the highly curved space–time close to the Sun. It also explains why unstable atomic nuclei accelerated to very high speeds in particle accelerators live longer than the same nuclei at rest. In addition, general relativity predicted the existence of gravitational mirages in the Universe fifty years before they were actually observed. These are distorted multiple images of distant objects caused by the fact that the light they emit is deflected when it passes through the highly curved space–time close to clusters of galaxies which actually lie in front of them. Such a cluster is said to act as a gravitational lens. Finally, taking an even more global view, our current cosmological picture was born of general relativity. Indeed, Einstein's equations already contain the seed for the models of the Universe studied by today's astrophysicists. The universe of general relativity, like the actual Universe we live in, is spontaneously expanding. Its space is dilating as time goes by from an initial singularity, the beginning of both space and time.

With his general theory of relativity, Einstein provided a coherent view of the Universe, showing that the laws of nature are the same for everyone, everywhere and throughout all time. In a cosmos where everything is in motion relative to everything else, Einstein tells us that there is no privileged observer, no absolute reference frame, only motions, masses, and energies, and relative distances and times, and a single absolute constant, the speed of light, absolute limit for all physical propagation. A time may come when the laws of nature will be unified at an even deeper level, in which some visionary physicist will discover that matter and energy are just compacted space–time, finally vindicating Albert Einstein's intuition that 'all is geometry'.

80 000 BILLION KILOMETRES OF NOTHINGNESS

Carried along by the gentle flow of air from Mir's droning air conditioning system, the apple

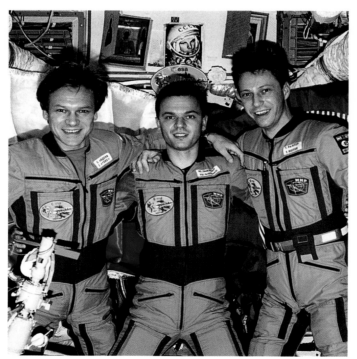

■ RUSSIAN COSMONAUTS SERGEI AVDEYEV AND YURI GIDZENKO POSE WITH GERMAN COSMONAUT THOMAS REITER BEFORE THE PORTRAIT OF YURI GAGARIN. SERGEI AVDEYEV IS THE MAN WHO HAS SPENT LONGEST IN SPACE, TOTALLING 748 DAYS DURING HIS THREE FLIGHTS IN EARTH ORBIT.

had drifted slightly. Sergei Avdeyev snapped out of his contemplation and took the fruit, biting into it cautiously. Through the porthole of the *Kvant* module, he watched the Earth slip away at 28 000 km/hr. Galileo and Einstein were right: motion is indeed nothing. Seen from up here, the Earth seemed a fragile place, enveloped within its blue shell, the atmosphere, clear as crystal. Here and there, huge burgeoning storm clouds were rising, dazzling white, as if they might burst right out of the skies. In reality, the atmosphere continues well above the clouds, growing ever more tenuous with altitude. In everyday, human terms, the Earth's atmosphere fades out above the highest peaks of the Himalayas. If Everest had been a few tens of metres higher than its actual 8 848 metres, Reinhold Messner would probably never have arrived at the top without an oxygen mask. However, there is still enough air for jet planes to counter gravity both through the upward thrust of air on their wings and the availability of oxidising agent for their reaction engines. Concorde flies at an altitude of 18 000 metres, whilst some fighter and spy planes climb to 30 or 40 kilometres. Even Mir, between 300 and 400 kilometres up, felt the slight resistance of the last few oxygen molecules in the exosphere. Beyond, of course, there is nothing. Or almost nothing.

So what is space? The space station was slipping through the Earth's night. For about three quarters of an hour, Mir would be plunged in darkness. Through the porthole, Sergei observed the stars. They shone with constant brightness, intense and bold against the absolute blackness of the velvety sky. Between the spacecraft, moving silently through the shadow of the Earth, and dazzling Sirius that gleamed like a diamond over there towards the south, there were 8.6 light-years of emptiness, more than 80 000 billion kilometres of nothing. An almost unthinkable distance, an impossible crossing, enough empty space to send anyone giddy. Sergei Avdeyev was looking out. There, just behind the bubble of air and life provided by the station, with its aluminium walls to protect it, a cosmonaut was approaching cautiously amongst the sparkling solar panel arrays. The unlikely space suit perched at the end of Strela, an articulated boom 20 metres long, was holding Sputnik 99 in his hand, a small radio emitter put together by and for radio hams. No one could be closer to the ultimate reality of the Universe than Jean-Pierre

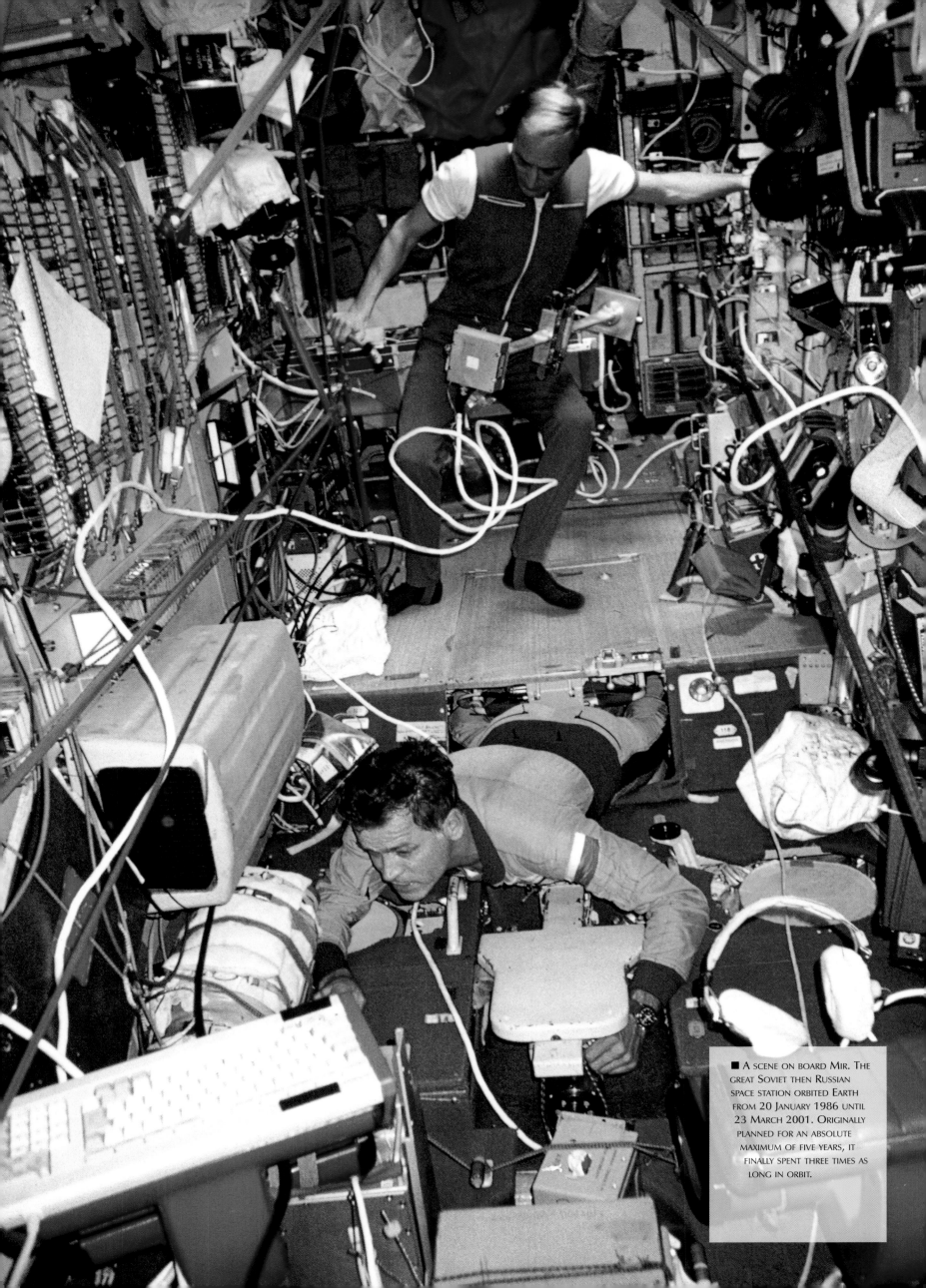

■ A SCENE ON BOARD MIR. THE GREAT SOVIET THEN RUSSIAN SPACE STATION ORBITED EARTH FROM 20 JANUARY 1986 UNTIL 23 MARCH 2001. ORIGINALLY PLANNED FOR AN ABSOLUTE MAXIMUM OF FIVE YEARS, IT FINALLY SPENT THREE TIMES AS LONG IN ORBIT.

■ SOME OF THE MIR MODULES
ACCUMULATED AN EXTRAORDINARY
COLLECTION OF JUNK DURING ITS
FIFTEEN YEARS OF OPERATION. IT
WITNESSED THE PASSAGE OF OVER
A HUNDRED COSMONAUTS FROM
ELEVEN DIFFERENT COUNTRIES.
THE VARIOUS CREWS LEFT
BEHIND THEM A GREAT
QUANTITY OF EQUIPMENT
AND THE REMAINS OF
MANY SCIENTIFIC
EXPERIMENTS.

Haigneré at this precise moment. He was going to walk in space for six hours, during which time he would orbit the Earth four times. He would symbolically carry out Newton's dream by throwing an object into orbit, just as in the idealised experiment described and even illustrated by the English physicist in 1702 in his *De mundi systemate liber*. Sergei watched his French colleague throwing Sputnik 99 into space. Naturally, the satellite did not follow the kind of parabolic trajectory we are so used to seeing when we fling a tennis ball into the air on the surface of our planet. Because it was freely falling, Mir represented a locally Galilean frame according to the general theory of relativity. Sputnik 99 thus began by adopting a roughly Galilean motion relative to it. In other words, for a certain period of time, it followed an approximately straight line as viewed from Mir. In the frame of reference of the Earth it was following an elliptical trajectory, tracing out a geodesic in Einstein's curved four-dimensional space–time.

ALBERT EINSTEIN'S CURVED SPACE–TIME

Space seemed completely empty around the cosmonaut. But what did this mean? In fact, at the dawn of the third millennium, physicists are unable to answer this question. Science still has many mysteries to solve in order to describe the medium in which Jean-Pierre Haigneré was floating, secured like an alpinist by a safety line only a few decimetres from Mir. To begin with, there was the ether, an invisible fluid imagined by nineteenth-century scientists to explain the propagation of electromagnetic waves. It is true that, with his relativity theory, Einstein eliminated this old-fashioned concept, quite useless to physics. However, he replaced it by a geometrical abstraction, four-dimensional space–time, a magnificent edifice but so difficult to grasp. A field unifying the world through gravitational waves, counterpart to the light waves of the electromagnetic field, discovered by Joseph Taylor and Russell Hulse in 1974. These American astronomers revealed the existence of gravitational waves emitted by a pair of extraordinarily dense and massive stars, the binary pulsar PSR 1913+16, in the heart of a profoundly curved region of space–time. It is hard to conceive of these stars, measuring no more than 10 kilometres across and yet more massive than the Sun. The density in their core is an almost unimaginable 10^{15} g/cm^3. What could a cubic centimetre of matter look like when it weighs a billion tonnes? And how could we describe the space–time associated with such bodies, so curved that it seems ready to close in on itself or slide indefinitely down its own slope?

The discovery of gravitational waves emitted by the binary pulsar was a genuine triumph for relativistic physics, and it won Taylor and Hulse the 1993 Nobel Prize for Physics. But what exactly is the geometric register of space–time? Contemporary physicists today seek to penetrate the smallest scales. Sometimes, where the shadows of general relativity and quantum mechanics overlap, they claim to have glimpsed the deeper secrets of the architecture of the Universe. In any case, the limiting scale for space–time is generally agreed today. Below 10^{-32} millimetres, they claim, there can exist nothing tangible. That is, at distances smaller than one hundred thousand billionth of a billionth of a billionth of a millimetre. According to certain opinions, space and time merge inextricably on this scale, perhaps swapping over, in a genuine spatio-temporal muddle where the principle of causality becomes inoperative and time itself may even flow backwards. On our own scale, space–time may well be smooth, curved and unruffled like the ocean viewed from the Mir cabin window, but on the subatomic scale, it becomes tumultuous and chaotic like the foam on the raging waves when a storm breaks over the wild seas.

For the time being, space seemed perfectly empty to Jean-Pierre Haigneré as he began to open the airlock and cross the threshold, and to Sergei Avdeyev who watched him from the atmospheric cocoon provided by the space station. It was of course only an illusion. The two cosmonauts knew perfectly well that space was not empty, and that without due care and attention, it could become lethal. At the altitude of Mir's orbit, space still contained several hundred thousand hydrogen, nitrogen and oxygen atoms per cubic centimetre. This may not seem very many when we recall that the air at atmospheric

■ PROGRESS SPACECRAFT REGULARLY RESUPPLIED THE MIR CREW WITH EQUIPMENT, OXYGEN, FOOD AND WATER. COSMONAUTS ALSO RECEIVED MAIL AND SMALL PRESENTS FROM THEIR FAMILIES.

temperature and pressure inside Mir contained more than a hundred billion billion atoms per cubic centimetre. However, from the astronomer's point of view, it is still very dense. In addition to this there is a prodigious quantity of light from the Sun. Billions of photons cross space each second. But perhaps most significant for the safety of cosmonauts in space are the violent bursts of protons expelled by magnetic flares on the solar surface. During these eruptions of high-energy particles, observed by Earth-based astronomers, cosmonauts must remain within the protection of the space station, or else risk receiving very high doses of radiation. There is also a continuous flow of elementary particles known as neutrinos. These are emitted in huge quantities from thermonuclear fusion reactions taking place in the Sun's core. Travelling at the speed of light, they are able to pass unhindered through the Earth, as though some magic were at work. For neutrinos, the Earth is like a perfectly transparent glass ball. A thousand billion of them were passing through Jean-Pierre, Sergei and Viktor each second, and yet they felt nothing. Jean-Pierre, well protected by his space suit, was immersed in a medium that was nevertheless not completely chilled.

■ THE EVER-CHANGING EARTH VIEWED FROM SPACE FASCINATED JEAN-PIERRE HAIGNERÉ DURING HIS TWO FLIGHTS ABOARD MIR. IN RARE MOMENTS OF FREEDOM, THE FRENCH COSMONAUT PHOTOGRAPHED BEAUTIFUL PANORAMAS OF THE BLUE PLANET.

According to physical theory, there is an absolute minimum temperature, another bound on nature like the speed of light, called zero kelvin, corresponding to a temperature of –273.15°C. Even in the absence of any solar radiation, and accounting for the heat emitted by the Earth, the Moon, the planets and the distant stars, if Jean-Pierre Haigneré had had an adequate thermometer to hand, he would still have measured a temperature of –270.42°C in space. Cosmologists call this residual radiation the cosmic microwave background. It is a relic of the earliest phases of the Universe when it was dense and hot. Just after the Big Bang some fifteen billion years ago, temperatures were extraordinarily high, but space–time has been expanding and cooling ever since.

Today, despite fifteen billion years of expansion since the Big Bang, space–time is still neither completely cold nor totally empty. If the cosmonauts on board the space station could make a trip across interstellar space, let us say to Sirius, they would cross ever more rarefied space. A few light-years from Earth, they would find something like ten atoms in every cubic centimetre of space. But in order to get some idea of the true space vacuum, a genuine foretaste of nothingness, the

cosmonauts must escape from our own galaxy, the Milky Way. They would have to move right out of the local cluster of galaxies that contains it and there, somewhere between Virgo and Coma Berenices, a hundred million light-years from Earth, naked space–time would finally be revealed in all its beauty.

In such a region as this, space is indeed essentially empty. On average, the intergalactic space crossed from time to time by a wandering photon contains only one atom per cubic metre. For the physicist, this means that the Universe has a density of the order of 10^{-30} g/cm^3, a thousand billion billion billion times lower than water. In the relativistic equations upon which cosmology is based, this figure corresponds to a Universe with hyperbolic curvature, implying that it is both infinite and eternal.

Jean-Pierre Haigneré had finished his space walk and closed the airlock behind him after 6 hours 32 minutes in the metallic science-fiction landscape outside, with the immense blue Earth above – or below – his head. The French cosmonaut and his two Russian colleagues would soon return to the planet of their birth. On 28 August

1999, Sergei Avdeyev would become the man who had spent longest in space. At the fabulous speed of 28 000 km/hr, he would have travelled more than 500 million kilometres around our planet, equivalent to a thousand times the distance from the Earth to the Moon. But did Sergei realise that, according to the special theory of relativity, during his 11 968 revolutions and 748 days orbiting around the Earth, he would have aged less than his wife and two children who remained at home? Of course, the gain was almost negligible. He would have economised a mere hundredth of a second! The fact is that, despite its extraordinary speed, Mir always remained under the dominion of Newton's laws of gravity, not so far removed from our everyday experience, and firmly attached to Earth and its denizens.

If relativistic effects are to become really significant, so that Newton must make way for Einstein, our spacecraft will have to travel a good deal faster. In fact, they must move almost infinitely fast compared to the speeds attained by today's rockets. Those adventurous travellers who seek one day to cross interstellar space must approach the chimerical horizon set by the speed of light.

■ WILL SPACE EVER BECOME A HOSPITABLE ENVIRONMENT FOR HUMAN BEINGS? NOTHING CAN BE LESS CERTAIN. OUTSIDE THE PROTECTIVE SCREEN PROVIDED BY OUR PLANET'S ATMOSPHERE, EVERYTHING IS HOSTILE AND DANGEROUS TO LIFE. SOLAR ULTRAVIOLET RADIATION IS BLINDING, WHILST THE SUN'S UNPREDICTABLE FLARES UNLEASH BLASTS OF INVISIBLE BUT EXTREMELY HIGH-ENERGY PARTICLES, EQUALLY LETHAL IN LARGE DOSES. THE GREATEST CHALLENGE CURRENTLY TAKEN UP BY TODAY'S SPACE EXPLORERS IS PRECISELY THE PROBLEM OF SHIELDING ASTRONAUTS AGAINST THESE SO-CALLED COSMIC RAYS.

Life beyond

■ AT THE BEGINNING OF THE YEAR 2001, ON THE FIFTEENTH ANNIVERSARY OF ITS ARRIVAL IN ORBIT, THE MIR SPACE STATION HAD REVOLVED ALMOST 100 000 TIMES AROUND THE EARTH, TOTTING UP OVER 4 BILLION KILOMETRES. THE RUSSIAN SPACE STATION WAS ALMOST CONTINUOUSLY OCCUPIED BY BETWEEN TWO AND FOUR ASTRONAUTS. THIS VERY FIRST SPACE VILLAGE OBTAINED ITS ENERGY SUPPLY FROM THE SUN.

■ THESE ORANGES AND GRAPEFRUIT FLOATING IN SPACE LIKE SO
MANY PLANETS WERE BROUGHT OUT AS A PRESENT BY AN AMERICAN
CREW VISITING THE MIR SPACE STATION.

The most noticeable thing was the silence. Once *Discovery* had rid itself of its mighty arsenal and torn itself from Earth's gravitational grasp in a furious clamour, it transformed into nothing other than a vast and elegant glider, streaking silently, high over the oceans, mountains and clouds of the blue planet. The space shuttle needed no energy to cruise through the upper realms at 28 000 km/hr. If it were not for the last few straggling molecules of the atmosphere rubbing gently against its streamlined fuselage, imperceptibly but inevitably slowing it down, it could go on revolving around the Earth indefinitely. The astronauts discovered the incredible change arising at the moment when the craft was injected into Earth orbit. Earlier, during the never-ending rehearsals they had endured in models and simulators at the Houston Space Center, there was a floor in the shuttle cockpit, and walls, and of course a ceiling. But now they found themselves within a quite undifferentiated enclosure, with neither top nor bottom. These notions no longer had meaning in space. And finally, there was the change of vista. The shuttle cockpit was a wonderful belvedere overlooking the Earth and the whole Universe. Above, below and behind the astronauts as determined relative to the way the pilot and flight commander's seats were orientated,

wide glass windows and portholes opened straight onto the cosmic panorama. Only 350 kilometres away, the Earth covered about half the field of view, a marvellous spectacle which astronauts never fail to remark upon when they return from a mission. Under the apparently stationary shuttle – or above it, depending on the mood of the astronaut – the blue planet rolled like a great noiseless ball across the darkness. The scene beneath the astronauts' gaze looked nothing like the photos downloaded from meteorological satellites. The latter orbit at 36 000 kilometres, from which standpoint our planet takes on the well-behaved and classical aspect of a simple globe that has been splattered with a few clouds like so much shaving foam. From the orbit occupied by the men and women travelling aboard Mir or the shuttle, and this includes all astronauts that have ever lived, with the notable exception of those twenty-four men lucky enough to have journeyed to the Moon, the perspective was truly unbelievable. It was rather like the scene that would be observed from an aeroplane flying at an altitude of 350 kilometres. Just beneath them, the cosmonauts could admire the terrestrial landscapes with startling clarity as they spun past. Flying over Paris, a French cosmonaut might identify the main thoroughfares of the French capital, whilst above Texas,

■ THIS MAN IS A LIVING
LEGEND. THE RUSSIAN DOCTOR
VALERI POLYAKOV HAD TWO
TRIPS ABOARD MIR. ON HIS
SECOND FLIGHT, HE SMASHED
THE RECORD LENGTH OF STAY
IN SPACE, REMAINING IN
EARTH ORBIT FOR OVER
FOURTEEN MONTHS

■ *DISCOVERY* SLOWLY APPROACHES THE MIR SPACE STATION. BEFORE BEGINNING THE FINAL DOCKING MANOEUVRE, THE TWO CRAFT MUST PRECISELY SYNCHRONISE THEIR ROTATIONAL SPEEDS. THE BRIGHT ORANGE

an American astronaut might recognise the beach where he or she has had a house built. In contrast, looking over towards the horizon, the view took on quite a different dimension. The seven islands of the Hawaiian archipelago could abruptly appear on the Pacific horizon, or the whole of Florida might suddenly stretch its long peninsula across the field of view, with Cuba and Puerto Rico trailing in the distance.

This enchanted vision of the Earth was continually changing as the planet rotated and the shuttle orbited around it. *Discovery* took about one and a half hours to complete one trip round the Earth. The sunlit hemisphere was thus crossed in forty-five minutes, whilst North America or Africa took only twenty minutes each! At this staggering rate, it was difficult for the astronauts to spend much time observing an idyllic atoll in the Pacific, a high-altitude lake in the Andes mountain range, or the motionless swell of the barchan dunes in the great Saharan dune fields. And what of the stars? It is true that they play a much more discreet role than planet Earth in the reminiscences of returned astronauts. But then, paradoxically, conditions are not ideal for observing stars from the shuttle. The portholes are too thick, lighting within the spacecraft is too bright, and in addition, there is barely time for human eyes to grow accustomed to the darkness – the night time only lasts forty-five minutes! All in all, the stars seem far less beautiful than they would viewed through the pure skies of an observatory high up a mountain on Earth.

LIKE A STAR ON THE HORIZON

Dominic Gorie, the pilot of the *Discovery* space shuttle, had nevertheless spotted a star on the dark terrestrial horizon, and it was a bright one, too. In fact, its brightness seemed to grow steadily. But unlike the stars that outlined the constellations and streaked across the sky at the same stunning speed as the Earth, it appeared to hang practically motionless in the sky, slightly above the blue line that marked the edge of the upper atmosphere. Very soon a small crowd had gathered behind the

pilot and the flight commander Charlie Precourt to watch in fascination a scene worthy of the film *2001, A Space Odyssey*. The whole crew of the shuttle was floating there: Americans Wendy Lawrence, Janet Kavandi and Franklin Chang-Diaz, and their 59-year-old Russian guest Valeri Ryumin, a space veteran who had already flown three times with Soviet comrades. The approaching star had soon swollen to a significant size and the crew could now admire the breathtakingly beautiful sight of Mir advancing. It was 4 June 1998 and this was the ninth and last rendezvous between the American space plane and the Russian outpost. Seen from here, the vessel was reminiscent of a great sailing ship with all its solar arrays deployed, roaming across the black and infinite abyss. The apparent fragility of the spacecraft, suspended between the immense ocean blue globe and ink black space, was poignant and instantly charmed the *Discovery* crew. Only Charlie Precourt and Dominic Gorie, fastened into their pilot seats, were completely engrossed in the delicate task of bringing the space glider towards the station.

Very soon they would dock with Mir at the hallucinatingly slow relative speed of a few centimetres per second.

The great space station, first Soviet then Russian, began to orbit Earth on 19 February 1986. Originally planned to run for five years at the very most, it went on to celebrate its fifteenth birthday in orbit on 19 February 2001 and was finally brought back to Earth after a total of 86 331 orbits on 23 March 2001. By then, it had welcomed more than a hundred cosmonauts on board, from a total of eleven different countries, and travelled more than four billion kilometres!

Viewed from the cockpit of the American shuttle, the scene appeared quite unreal. The shuttle loading bay had opened to reveal the docking airlock shortly to mate with its counterpart on Mir. The station, now just 'above' the shuttle, folded back its huge solar arrays, thus streamlining somewhat its austere and imposing silhouette. The *Discovery* astronauts watched in awe. Unlike the American space shuttle, designed to operate both in the Earth's atmosphere and in empty space, and which could be admired as it reached up from its launch pad just before each

■ THE DISCOVERY CREW APPROACH THE PROUD RUSSIAN SPACESHIP WITH SOME EMOTION AS IT ADVANCES ACROSS SPACE, SEEMINGLY PROPELLED BY THE GENTLE SOLAR BREEZE. IN HALF AN HOUR OR SO,

flight, nobody on Earth could even begin to imagine the impression made by the size and complexity of Mir. Naturally, it was designed at the outset to operate in space alone and took full advantage of the laws of microgravity. The various modules that made it up splayed out in all directions. Over the years, Mir had been assembled in orbit and no life-size model of it existed anywhere on Earth. In any case, no full-scale model on Earth would be able to simulate it. Astronauts like those arriving on board *Discovery*, and preparing to visit Mir for the first time, would thus discover it in the true sense of the term. Mir was indeed the very first human habitation designed for space and only for space. It was perhaps the first step on the long road to the stars. Only time will tell.

Even pressing their faces against the shuttle portholes and contorting themselves to obtain the best possible view, the astronauts would be unable to appreciate Mir in all its majesty. They would see before them only a part of the vast space-built

Meccano set that formed a kind of three-dimensional cross in the void. At the heart of the whole ensemble was a large module which the astronauts would have some difficulty in making out, so many were the appendages growing out from it. This was the original Mir, placed in orbit in February 1986. It already constituted a space station in its own right. The 13-metre long cylinder had a diameter of 4.2 metres and weighed 20 tonnes. It was the nerve centre of the great spacecraft, and the living area, containing both kitchen and restaurant, a genuine meeting place for cosmonauts who sometimes spent the whole day in the wandering passageways of the station. Mir was equipped with two docking units, one at each end, allowing it to combine with a total of six further modules. In April 1987, a second module was launched from Baikonur by a Proton rocket and guided automatically towards Mir. It docked directly onto the end of the first module, extending it by 6 metres and increasing its mass by 6 tonnes. In December 1989, the large *Kvant 2*

module, 13.7 metres long, 4.3 metres in diameter and weighing almost 20 tonnes, docked at right angles to Mir. The little space-based assembly was then joined by *Kristall* in June 1990, docking onto the end of *Kvant 2* and in line with it. At this point, the whole game grew more complex as the real space Meccano construction got under way. At the beginning of 1995, the Russian cosmonauts moved *Kristall* in order to make way for the next module to leave for space. They isolated the *Kristall* module from Mir, detaching it and displacing it along the arc of a circle around the station. They then docked it once more, but perpendicular to its original position. It is worth mentioning that *Kristall* measured 12 metres long and weighed almost 20 tonnes. It was equipped with an impressive system of retractable solar panels with a wingspan of 27 metres. Having relinquished its place on Mir's main

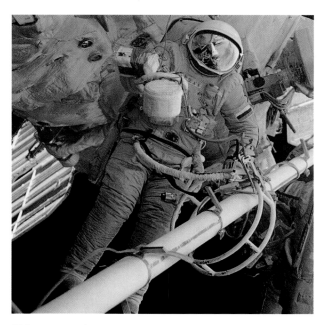

■ LIKE DEEP SEA DIVING, EXTRA-VEHICULAR ACTIVITIES ARE EXTREMELY DANGEROUS. WHEN THEY RE-ENTER THE VESSEL, COSMONAUTS ARE EXHAUSTED.

docking port, *Kristall* was replaced by *Spektr* in June 1995. Like most of the modules launched by the Proton rocket system, *Spektr* weighed 20 tonnes. It had a diameter of 4.3 metres and a length of 13.7 metres. Finally, in April 1996, the last module of the Russian space station was docked as an extension to *Kristall*. This was *Priroda*, equipped like the other modules with a vast system of solar panels. The station could thus satisfy all its energy requirements, with the total available power from the whole system reaching 34 kW. A lot of electricity was used on board to heat and light the station and to operate various pieces of equipment needed to survive in space, in particular, the production of oxygen and water. Needless to say, the many scientific experiments running on board also required an energy supply.

Three docking ports remained available: one on *Kristall* where the American space shuttle would link

up, one at the end of the *Kvant* module, and a third on Mir itself. The last two were designed to link with manned Soyuz craft arriving from Earth and the small Progress resupply vessels. For safety reasons, a Soyuz was permanently docked onto the station, ready to escape back to Earth with all Mir's inhabitants should the need arise.

In all, the Mir space station thus comprised six inhabitable modules, plus the Progress and Soyuz spacecraft that regularly came to join it. In this configuration it reached a total length of 45 metres and a maximum mass of 110 tonnes! But it was the 400 cubic metres of inhabitable volume that was really impressive, equivalent to the volume of a large six-room flat back on Earth. The big difference, of course, is that on Earth we are mainly concerned with the inhabitable area, whereas in the six rooms of Mir, the cosmonauts could equally well sleep on the walls

■ THE COSMONAUTS NEVER SUCCEEDED IN REPAIRING THE OXYGEN LEAK IN THE *SPEKTR* MODULE. CONDEMNED, IT WAS DELIVERED UP TO THE VOID.

and ceiling. They never walked on anything they might call a floor. Indeed, the word seems to lose all its meaning to those living the experience of weightlessness. Cosmonauts were quite used to floating from module to module, taking care not to bump into a colleague arriving head first in the opposite direction through Mir's main docking port.

This risk obviously increased when Mir had visitors aboard, as in the case of this last rendezvous with the American shuttle. Indeed, ten men and women would share these heavenly apartments for the next few days. And nothing could be further removed from a typical American residence than this Russian dacha. Out in space, the deep cultural divide that separates the two great nations is even further amplified. On the American side, there is a genuine obsession with tidiness and cleanliness. Nothing is left to chance. In a word, an American shuttle must

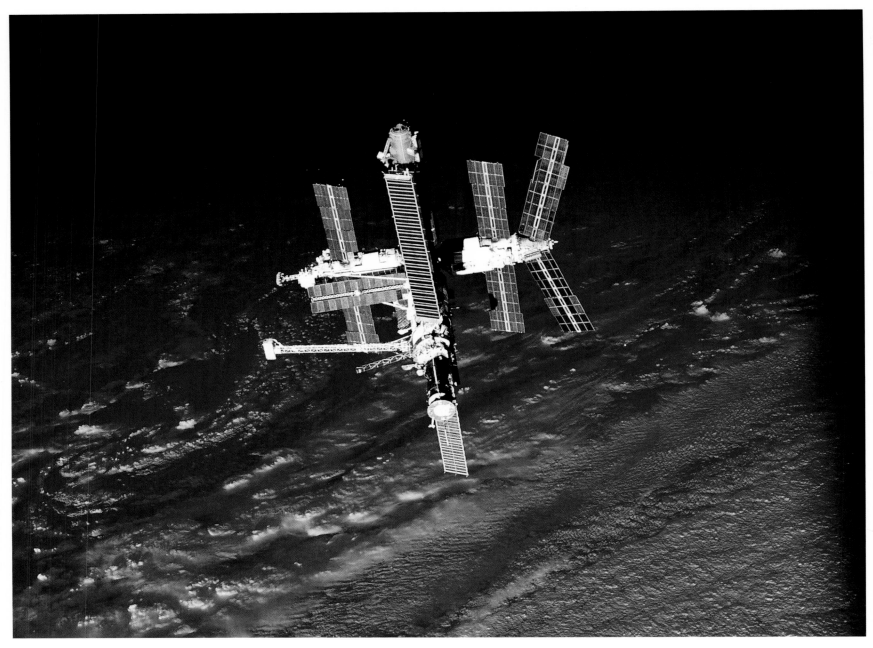

■ WITH THEIR SPACE STATION MIR, THE RUSSIANS ACQUIRED INVALUABLE EXPERIENCE IN SPACE FLIGHT. SERGEI AVDEYEV TOTALLED 748 DAYS IN ORBIT, VALERI POLYAKOV 678 DAYS IN JUST TWO FLIGHTS, AND ANATOLIY SOLOVYOV 651 DAYS IN FIVE FLIGHTS.

be spotless. This obsession is managed more easily by American technicians and astronauts in that each shuttle is completely overhauled between two flights. In addition, shuttle missions are very short, never exceeding two weeks.

MIR: FIFTEEN YEARS OF FAITHFUL SERVICE

The *Discovery* astronauts explored Mir, wandering from module to module. It was the very antithesis of the shuttle they had just left. Over the years, cosmonauts occupying the huge and ageing space station, sometimes for periods of six months or a year, had naturally made themselves feel at home the Russian way! With each breakdown, inevitable in such a gigantic chunk of high technology so far beyond its nominal life expectancy, successive cosmonauts had implemented a treasury of instant repairs, whilst living conditions had gradually deviated further and further from astronautic standards. Computer failures, breakdowns in air-conditioning and oxygen production, radio communication problems, and difficulties with pressurisation of manned modules had now become commonplace aboard Mir. In places, damp had begun to eat into electrical cables, whilst in others small pools of water had accumulated in the vicinity of sensitive electronic equipment. Elsewhere, dust and mould polluted the already stuffy environment of the station. After more than ten years of continuous occupation by over 100 cosmonauts, Mir was an unbelievable shambles, piled high with unusable computers and layer upon layer of sophisticated equipment whose function no one could now remember. Every imaginable kind of tube and cable hung along the walls of the modules, effectively obstructing the safety airlocks, designed in principle for rapid closure if pressurisation problems should occur aboard. In fact, no one seriously considered such a disaster scenario. At least, not until February 1997.

Continuous occupation of space has been the dream of scientists and engineers since the advent of space travel. In the 1960s, the pioneering flights of Yuri Gagarin, Alan Shepard, John Glenn, Gherman Titov or Valentina Tereshkova were made in tiny capsules where they remained stretched out whilst the vehicle hurled itself round a few orbits. When the record was shattered by Titov in August 1961, with twenty-four hours spent in space, this was considered by all as an astonishing feat. Later, for the long crossings between the Earth and the Moon, living conditions remained spartan aboard Apollo vehicles. Nobody would have dared to ask astronauts to engage upon longer periods in space whilst modules were limited to a few cubic metres of living space. The solution arose quite naturally when

the American lunar exploration programme came to a premature close with the cancellation of Apollo 18, 19 and 20 for reasons of economy and safety. NASA suddenly found itself with a fleet of giant Saturn V launchers and nowhere to send them. The American engineers then had the idea of using one of these superpowerful rockets to carry a sizeable manned station into Earth orbit, and so the Skylab project was born. In a context of drastic budgetary cutbacks, NASA engineers were compelled to bring their imagination to bear. Skylab was quite simply manufactured out of the third stage of the rocket. Indeed, this stage had become totally superfluous for the purposes of reaching an Earth orbit, a small leap for such a vehicle. The gigantic tank, originally intended for liquid oxygen and hydrogen, was pressurised and made inhabitable. It was then equipped with various medical and scientific laboratories, together with a high-performance solar telescope. The last Saturn V placed it in orbit on 14 May 1973. However, its solar and meteoroid shield was torn off during ascent. This removed one solar panel wing, whilst the other was jammed. The temperature in the station rose to 52°C and the crew could not be launched until procedures had been developed to render it inhabitable. After a ten-day delay, on 25 May of the same year, the first crew arrived aboard the crippled station. They and subsequent crews managed to meet virtually all mission objectives in an unprecedented salvage operation that has never since been repeated. The first crew, Charles Conrad, Paul Weitz and Joseph Kerwin, were sent into orbit aboard an Apollo module launched by the much less powerful Saturn 1B rocket. They stayed in orbit for twenty-eight days, and once the main repairs had been made were able to enjoy the comfort of the spacious Skylab station with its unbelievable 320 cubic metres of living space, a novelty in those early days. The three astronauts felt as though they were acting out a science-fiction film. Charles Conrad in particular had a clear recollection of the long days spent with his two fellow astronauts in the exiguous six cubic metres of the Apollo 12 command module. A second crew replaced the three men on 28 July 1973. This time Alan Bean, Jack Lousma and Owen Garriot spent almost sixty days in orbit. Then, on 16 November 1973, Gerald Carr, William Pogue and Edward Gibson set up house for a final mission that was to smash the space occupation record of the day, remaining in orbit for eighty-four days.

A completely unknown universe was revealed to the nine Skylab astronauts. This was the world of weightlessness as an exploitable medium. In the roomy space station, all kinds of experiment and wild capers could be catered for! But for the doctors back in the Houston control centre whose task it was to monitor the astronauts' behaviour, this was a big step into the unknown. Would the absence of weight lead in the long term to serious physical problems? Would the metabolism of these men suffer permanent modification? Finally, and above all, everyone was wondering whether human beings were really made for life in space.

THE RUSSIAN SPACE EPIC

The first Skylab crew spent almost a month in space. This was slightly longer than the twenty-four days the Soviet crew (Dobrovolsky, Patsayev and Volkov) had managed to endure aboard their small Salyut 1 space station in 1971. Salyut 1 was the world's first manned space station. Its success was marred when the entire crew perished on re-entry due to pressure failure. When the first Skylab crew returned to Earth, they were exhausted and their physical condition seriously worried the doctors. Their health had clearly deteriorated. Weakened by muscle tissue loss and alarming calcium deficiencies that posed a considerable threat to their bone structure, weakened further by reduction in red blood cells, and perturbed by the novel redistribution of fluids within the body, these men clearly could not have lived much longer in microgravity conditions. The doctors thus invented physical exercises for the astronauts participating in the last two Skylab missions, and despite the fact that they remained two or three times as long in space, they returned to Earth in much better form.

After the record sojourn of eighty-four days in space, established by Gerald Carr, William Pogue and Edward Gibson, the Americans strangely abandoned the development and study of long space flights, devoting themselves entirely to the costly enterprise of the space shuttle. They left Skylab in orbit in the hope that the first shuttle flight might link up with it again in the late 1970s. However, this spectacular rendezvous was never to be. The American shuttle programme was delayed and the scant air molecules of the upper atmosphere eventually got the better of Skylab, slowly but inexorably reducing its speed and dragging its trajectory down towards the surface. On 11 July 1979, Skylab entered the atmosphere and abruptly

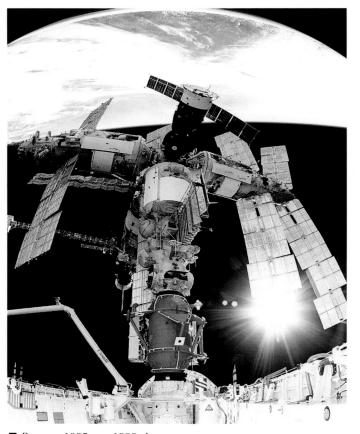

■ BETWEEN 1995 AND 1998, AMERICAN SHUTTLES LINKED UP NINE TIMES WITH THE MIR SPACE STATION IN EARTH ORBIT. THIS WAS THE PERFECT OPPORTUNITY FOR AMERICAN ASTRONAUTS TO TAKE PART IN LONG SPACE FLIGHTS.

disintegrated, significant pieces reaching the ground in Australia.

Whilst the Americans were proving conspicuous by their absence, sending no astronauts into space between 1975 and 1981, the Soviets took up the challenge, achieving new levels of know-how in the domain of long human space flights. Following the disastrous failure of their plans for lunar conquest, the Soviets turned their attention to human flight towards Mars. Because of the great distances involved, about 500 times the distance from the Earth to the Moon, this obviously meant spending several months or even years crossing interplanetary space. When the Americans abandoned Skylab, the Russians were already proceeding with their own space station programme. Salyut 2 had been launched in April 1973, but its engine exploded and it went out of control after only two weeks. Luckily there was no crew aboard. In 1974, an improved generation of space stations was put into orbit. This was Salyut 3 and 4, followed by Salyut 5 in June 1976. Equipped with a single docking system and designed to link up with a manned Soyuz spacecraft, these stations could only be visited by a single crew of two cosmonauts and could not be resupplied in flight. After their disintegration in the Earth's atmosphere, they were soon replaced in September 1977 by a new station, Salyut 6. Measuring 13 metres long and weighing 20 tonnes, this robust station with a docking port at each end would allow Soviet scientists to acquire a considerable level of expertise in space link-ups. Over five years, twenty-seven cosmonauts and twenty or so unmanned spacecraft were to visit the new station. With both ports occupied, Salyut 6 was 30 metres long and weighed in at 33 tonnes. Inside, the crew enjoyed a living space of almost 100 cubic metres. The great Salyut adventure reached a climax with Salyut 7, launched in April 1982 and abandoned in 1986 in favour of the first Mir module. It burnt up in the atmosphere in February 1991. Cosmonauts accomplished many feats on board Salyut 6 and 7, indulging in regular space walks and shattering the American record for the longest time spent in space, held until then by Skylab astronauts. To begin with, the record was taken to three months, then four, and then six. In 1984, the crew composed of Leonid Kizim, Oleg Atkov and Vladimir Solovyov remained for

A TRUE SCIENCE-FICTION
SCENE, WORTHY OF THE FILM
2001, A SPACE ODYSSEY.
ATLANTIS LINKS UP WITH THE
RUSSIAN SPACE STATION.
PHOTOGRAPHED BY
ANATOLIY SOLOVYOV
ABOARD A SOYUZ CRAFT
SPECIALLY RELEASED FROM
MIR FOR THE OCCASION.

237 days aboard Salyut 7, almost eight months in orbit. Now no one could ask whether humans were able to survive in space. Cosmonauts were better and better monitored by Soviet, then Russian, doctors, engaging in regular physical training during the flight in order to maintain muscle tissue. A great many cardiovascular experiments were carried out both in flight and after return to Earth. In view of their great physical condition, cosmonauts were no longer satisfied with spending a few months in space, but sought to return for further visits. It was probably at this point that the human space adventure really began, when the Soviets began to assemble the components of the Mir space station with such exemplary perseverance and serenity. Owing to the small living space in their space shuttle and its limited orbital lifetime, the Americans were completely excluded from this particular adventure. All those who have made long space flights, aboard Salyut 6 and 7, and then aboard Mir, are Russian.

Over three flights, Sergei Avdeyev accumulated 748 days, or more than two years, in orbit. Then the extraordinary doctor Valeri Polyakov totalled 678 days in just two flights, whilst the intrepid Anatoliy Solovyov flew for 651 days during five different flights. Dozens of Russian cosmonauts have flown aboard the three stations, a few weeks or months at a time, gradually building up a wealth of data for the medical and biological sciences. The first non-Russian cosmonaut to join the ranks of these was not an American at all, but a Frenchman. In 1999, Jean-Pierre Haigneré spent six months aboard Mir, beating American space veteran Shannon Lucid by a few hours. The latter, aged fifty-seven in the year 2000, has flown on five space

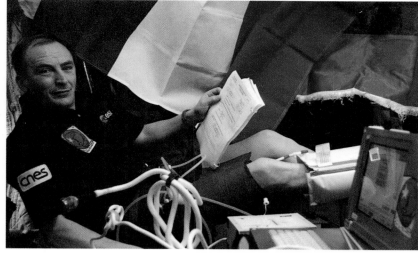

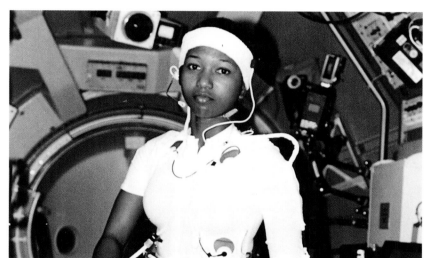

shuttles but had to accept a Russian invitation to join the Mir crew in order finally to achieve a long endurance record.

THE WONDERFUL JOURNEY OF VALERI POLYAKOV

Of all the cosmonauts to spend time aboard Mir, the most endearing is undoubtedly Valeri Polyakov, with his singularly unusual career. Born in Toula, Russia, in 1942, this doctor was selected for the Soviet cosmonaut corps at the age of thirty. While training at Star City in Moscow, he was first chosen as a reserve for the Soyuz T3 flight, launched in November 1980, but was not called. Once again, in 1984, he was a member of a back-up crew, and patiently played out his role as substitute. Finally,

on 29 August 1988, now aged forty-six, he left for space aboard the faithful Semiorka. On his first trip aboard Mir, he stayed almost eight months. On 8 January 1994, at the age of fifty-two, he set off for Mir once again and made the longest space flight in history, remaining 438 days, more than fourteen months, in orbit. The Russian doctor greatly enjoyed this extraordinary experience. Through the station's portholes, he was able to admire more than 7000 sunrises and sunsets above the clouds of the Earth's atmosphere, cross the oceans at unthinkable speed, fly over the five continents, and observe the subtle changes in the face of the blue planet as the seasons went by. After Polyakov's two flights, covering twenty-two months in space, scientists at last had the proof that human beings could indeed adapt to life in space and that, if they desired, they could one day detach themselves from the bonds of Earth's gravity.

■ The unbelievable view enjoyed by astronauts from the porthole of the *Atlantis* space shuttle when docked to Mir. The Andes mountain range slips gently beneath the silvery wings of the space station. Lake Titicaca is clearly visible on the right, just under a solar panel, on the border between Bolivia and Peru. Reaching out above the same panel are the Strela boom arm and spidery Sofora mast structure where cosmonauts experience their most dizzying sensations.

■ AFTER THE LUNAR CONQUEST, THE AMERICANS USED THE LAST STAGE
OF A SATURN V ROCKET TO CREATE A LARGE SPACE STATION. THE HUGE
SKYLAB HOUSED THREE CONSECUTIVE CREWS BEFORE BURNING UP IN THE
EARTH'S ATMOSPHERE.

Cosmonauts engaged in many scientific experiments aboard Mir, just as they had aboard Salyut and Skylab, and continue to do so aboard the shuttle. The latter often carries laboratories in its cargo bay, such as Spacelab, designed by the European Space Agency (ESA), and the American Spacehab. Clearly, the most important experiments concern biology, physiology and space medicine. Having convinced themselves by the end of the 1970s that long-term human space flight in weightless conditions was actually possible, American and Soviet doctors and biologists are now seeking a better understanding of the physiological changes that take place under these conditions. In particular, they are searching for ways to parry the more serious symptoms, such as the notorious space sickness which affects more than fifty per cent of cosmonauts to varying degrees. This extremely handicapping problem is due to changes in the working of the vestibular organ, located in the inner ear. On Earth, this organ is part of our balance system. It takes the vertical as reference, that is, it refers to the direction in which the weight is felt to pull wherever we happen to be situated. In space, however, the brain receives incomprehensible information and manifests its confusion through various disorders such as dizziness. Needless to say, these greatly reduce astronauts' ability to work correctly and impair their quality of life. Apart from trying to improve the well-being of the crew during missions and upon their return to Earth, the scientists' aim is to provide easier access to space for a growing community of human beings. In the days when astronauts had to be made of the stuff of heroes, they were selected on the basis of stringent physical and psychological criteria, both in the Soviet Union and the USA. In reality, only fighter pilots could ever hope to fit the bill. These men were in excellent physical condition and were used to enormous load factors of four, five or six times the acceleration due to gravity when flying military aircraft. At ease in the tiny cockpit of sophisticated and dangerous machines, capable of carrying out all sorts of acrobatics in the air without ever feeling the slightest discomfort, they could soon adapt to the kind of conditions prevailing in space. But once the work of the pioneers had been accomplished, reaching Earth orbit and then walking on the Moon, scientists hoped that space would become a genuine laboratory for fundamental research.

■ THE AMERICAN SHUTTLE IS FLEXIBLE TO USE AND REMARKABLY AGILE IN SPACE, QUALITIES EXPLOITED ON SEVERAL OCCASIONS TO RECOVER AND SAVE SATELLITES. HERE ASTRONAUTS PREPARE TO CAPTURE THE HUBBLE SPACE TELESCOPE WITH THE REMOTE MANIPULATOR ARM.

SCIENTISTS IN SPACE

In the 1980s, thanks to progress made in space medicine, the first civilian research scientists were able to take their place aboard space shuttles and Semiorka rockets. These were men and women who had never set eyes upon the joystick of a fighter plane. They included doctors, in particular, such as Valeri Polyakov, and Claudie André-Deshays who undertook a sixteen-day mission aboard Mir in 1996 and is currently preparing for other space flights. This French astronaut carried out many biomedical experiments during the Cassiopeia mission, mainly extensions of the space science programme laid out by the French Centre national d'études spatiales (CNES). Today, France is the country that has most heavily invested in the conquest of space after the United States and Russia. As of the year 2000, bilateral agreements with the two great space powers have made it possible for eight French astronauts to take part in fourteen missions, either aboard US shuttles or on Russian spacecraft. As early as 1982, Jean-Loup Chrétien flew out from Baikonur to the Salyut 7 space station. He then took part in the Aragatz mission aboard Mir in 1988, this time in the company of a Russian-

American crew aboard *Atlantis*! The French cosmonaut was sixty-two in the year 2000 and is still in active service. Between 1992 and 1999, further missions to Mir were Antares, Altair, Cassiopeia, Pegasus and Perseus, during which Michel Tognini, Jean-Pierre Haigneré, Claudie André-Deshays and Léopold Eyharts helped to push forward a new medical discipline known as gravitational physiology. For their part, Patrick Baudry, Jean-François Clervoy, Jean-Jacques Favier and once again Michel Tognini set off from Cape Canaveral aboard *Discovery*, *Atlantis* and *Columbia*.

Jean-Jacques Favier perfectly prefigures the new generation of astronauts. He is primarily a research scientist, with a doctorate in physics and specialising in spaceborne metallurgy and crystallogenesis. Unlike the majority of his precursors, he did not go into space to perfect the art and practice of space flight with all its associated safety problems, but rather to contribute to scientific progress. The new astronauts, doctors, chemists, physicists and others, came onto the scene at a crucial moment during the 1990s, a time when doubt had begun to pervade the minds of politicians, decision-makers, financial backers and the general public. Questions were being asked about the cost and the

■ SOMEWHERE IN EARTH
ORBIT, DECEMBER 1999.
ASTRONAUTS GENTLY LOAD
HUBBLE INTO *DISCOVERY'S*
CARGO BAY. MANY SPACE
WALKS WILL BE NEEDED TO
REPAIR THE SPACE TELESCOPE
AND CHANGE ITS
ONBOARD COMPUTERS.

practical advantages of a human presence in space. Indeed, with the hindsight provided by four decades of space exploration, it was not difficult to see that the main part of the enormous progress witnessed over this period had been accomplished without any human participation!

In fact, at the beginning of the twenty-first century, space has undeniably come back down to Earth. It is now present throughout our everyday lives, to the extent that each day we regularly use space technology without even realising it. Perhaps telecommunications provides the best example. Over the last thirty years, dozens of unmanned satellites have been put into orbit to relay telephone calls and radio and television broadcasts around the planet. Satellites are also responsible for the development of portable telephones, not to mention the ubiquitous GPS (Global Positioning System) which, widely used in aviation and shipping, can now be accessed from private cars. Another area is weather forecasting. Meteorological satellites can monitor the weather around the globe in real time, giving more reliable predictions and tracking major features like cyclones. Still other satellites like SPOT or ERS map the continents, study earthquakes, monitor agriculture and deforestation, measure sea levels and the movements of ice fields, and in short carry out global ecological surveillance of the whole planet. All these devices occupy orbits at much higher altitudes than can be attained by either the American shuttles or the Russian Semiorka rockets. The shuttle is occasionally sent up to altitudes of 600 kilometres, whereas unmanned rockets, such as Russia's Proton, Europe's Ariane 4 and 5, and America's Titan and Delta rise lightly up to the useful orbits, generally situated between 800 and 36 000 kilometres. The latter is the so-called geo-stationary orbit, particularly sought after by telecom-munications operators and weather institutes. From this spectacular viewing point, satellites revolve about the Earth with the same period as the Earth's rotation itself, so that the same point on the surface always lies below the satellite's observing equipment. In this way a satellite like Meteosat can be stationed precisely above Europe, whilst others can be located above North

■ SPACE WALK ASTRONAUTS CARL WALZ AND JAMES NEWMAN SEEM TO BE FLOATING ABOVE THE CARIBBEAN. JUST BEYOND THEM CAN BE SEEN THE TILES THAT PROTECT THE *DISCOVERY* FUSELAGE.

America, South America or Asia. In contrast, most spy satellites are launched into highly elliptical orbits from top secret military bases. This allows them to fly past certain regions on the surface at a mere 100 km altitude during each revolution of the planet. From such distances, the most powerful Russian and American spy satellites, equipped with telescopes several metres across, are able to resolve details down to a few centimetres. Such accuracy does not come without its problems. These satellites are soon slowed down by the upper layers of the Earth's atmosphere and have lifetimes of the order of a few weeks at best.

Exploration of the Solar System, one of the greatest achievements in the history of mankind, has also been carried out by unmanned spacecraft. The eight major planets, fifty or so of their natural satellites, several asteroids and even comets have been subjected to close examination, observed from various angles and analysed chemically by genuine spaceborne robots. Operating by remote control on planets such as Venus and Mars, they have led to extraordinary advances in our knowledge of the history and geography of the Solar System. And what of the Moon? the reader may ask. Who remembers today that, after the exploits of the American astronauts, Soviet scientists also visited our natural satellite with the help of robots, exploring a part of its surface with their little unmanned lunar rover Lunakhod? Lunar probes Luna 16, 20 and 24 even brought back precious samples of lunar rock, a clear scientific proof of lunar conquest. In fact, in every area of space exploration, the presence of human beings has never proved to be crucial. This is borne out by the fact that the long cherished hope that human beings could create new molecules for the pharmaceutical industry or revolutionary new materials in the microgravity conditions prevailing aboard space stations has up to now proved elusive. Worse still, it was soon observed that the weightless conditions sought by research laboratories were never quite achieved during manned flights. The erratic motions of the astronauts themselves perturbed the experiments. And as often as not, these experiments could just as well have been carried out by robots.

However, the debate about whether a human presence in

space is really useful should probably never have taken place. For one thing, it is a considerable technological and human challenge to send astronauts into space and return them to Earth alive, and this in itself constitutes a source of inspiration and a sign of scientific excellence for those countries that succeed in such a venture. On a more basic level, it can be argued that the conquest of space needs no practical or economic justification. It is a question of ideals, utopia, instinctive curiosity, an indefinable metaphysical feature of human beings that compels them, maybe genetically, to explore any new horizon that presents itself. Indeed, scientific objectives are no longer emphasised by the great nations that participated aboard Mir and today are setting up the International Space Station. Instead, they stress the human adventure, the need to explore and the desire for cooperation between different cultures. So it matters little how many thousands of billions of euros are spent so that twelve men can walk on the Moon, or 400 men and women can float in weightless conditions, provided that they involve the whole of humanity in their adventure, by proxy as it were. Through their eyes, we may admire the Earth from a little higher, appreciating that we all share the same unique and fragile world.

From Clément Ader to John Young

The flight of the space shuttle is a pure and idealised reminder of the very adventure of aviation itself. A hundred and ten years ago, when the extraordinary perseverance of French engineer Clément Ader finally succeeded in getting the first aeroplane off the ground, nobody would have imagined that *Éole* was the precursor of Concorde and Airbus. Like Ader, Santos-Dumont or Blériot, twenty-first-century astronauts are totally devoted to the

flight of their complex and fragile space vehicle. They play out once again the difficult role of the pioneer. From this standpoint, the American shuttle or the space station Mir are subject to precisely the same constraints. Ninety per cent of the years of training undertaken by men and women destined for space involves maintenance and safety for the crew during lift-off, orbit and return to Earth. In these conditions, it is quite understandable that there should be such limited scope for scientific endeavour during the mission. The fruits may come one day, but much later, and nobody can say today what harvest future astronauts will gather.

Whatever happens, they will have written one of humanity's great sagas, with some episodes of epic proportions. We now know, thanks to the investigations of American journalist Bryan Burrough, that the space odyssey almost met with a second terrible disaster in 1997, this time aboard Mir. Indeed, it might

well have set back mankind's quest to reach the heavens by many years.

No longer able to finance its maintenance in orbit, the Russians eventually transformed Mir into a kind of luxury space hotel for western astronauts. But over several months in 1997, the Russian and American cosmonauts had other things to contend with than the scientific experiments entrusted to them by ground-based research centres. In his behind-the-scenes account *Dragonfly: NASA and the crisis aboard Mir*, Burrough reveals that in that particular year the spacecraft and its occupants were exposed to great danger on two separate occasions.

The first incident happened on 23 February, when fire broke out aboard the station. Cosmonaut Aleksandr Lazutkin was floating unhurriedly around the *Kvant* module preparing to change a canister of lithium perchlorate for the oxygen generator. Suddenly, he saw a long flame reaching out from the device. His first reaction was to smother it with a damp towel, but this failed to have the desired effect. Seeing the hissing flame grow still longer, he called out to his fellow cosmonauts as calmly as he could: 'Hey, you guys, there's a fire.' His four companions, the German Reinhold Ewald and the Russians Vasili Tsibliyev, Aleksandr Kaleri and Valeri Korzun came rushing to his aid, grabbing the nearest fire extinguishers as they went past. Inside the *Kvant* module, apart from the fire teased by oxygen liberated from the canister, the smoke was beginning to grow pungent. The flight commander, Korzun, ordered the crew to put on their oxygen masks. At this juncture, the cosmonauts ran into a series of problems that seriously worsened the situation. To begin with, one of the extinguishers was not working, whilst two others could not be unhooked from the wall. In the confusion, Korzun realised that they had forgotten about the sixth cosmonaut then travelling with them aboard Mir, the American Jerry Linenger, who was at that very moment settling down to sleep in the *Spektr* module. Five men were now struggling together to put the fire out, whilst Aleksandr Lazutkin began to prepare one of the two Soyuz craft docked onto Mir for an emergency evacuation. The other Soyuz, docked onto the far end of the *Kvant* module, could not be reached without crossing the fire. However, Korzun finally managed to put the fire out. Standing in the darkness, his face hidden by the oxygen mask that prevented him from breathing in the toxic fumes, the commander finally took control of the situation. But it had been a narrow escape. Although officially, according to reports put out by NASA and the Russian space agency, the incident lasted a mere ninety seconds, the cosmonauts admitted when they returned to Earth that they had fought the fire for a good ten minutes in the confined space of the station as it hurtled through the void at 28 000 km/hr.

As if this were not sufficient, a second incident made 1997 a truly terrible year for Mir. On 25 June, exhausted after many stressful months spent in a space station that was beginning to behave in unexpected ways, Vassili Tsibliyev took the controls of the TORU guidance system that would allow him to dock the Progress M34 craft onto the station. Using the two levers of the new manual docking system, he could orientate Progress by remote control. It was

■ JAMES NEWMAN EXPLORES THE CARGO BAY OF THE SPACE SHUTTLE, SECURED TO THE WALLS LIKE A MOUNTAINEER. WITHOUT SUCH PRECAUTIONS, ASTRONAUTS WOULD RUN THE RISK OF FLOATING AWAY FROM THE SHUTTLE AND GETTING LOST IN SPACE FOR EVER.

rather as though he were actually aboard the little vessel, a camera fixed onto the front providing permanent coverage of Mir as seen from Progress. The small resupply vessel had left Baikonur two days previously and now Mir appeared as a bright star on Tsibliyev's screen. Applying a thrust via its engines, Tsibliyev brought the craft towards Mir at an almost imperceptibly slow rate. The whole crew was agog as the delicate approach manoeuvre got under way. Lazutkin was still aboard, but the American Jerry Linenger had been replaced by Michael Foale, whilst Reinhold Ewald, Valeri Korzun and Aleksandr Kaleri had returned to Earth with Linenger. With his face pressed against one of the portholes, Foale was using a laser ranging device to obtain exact measurements of the velocity and distance of Progress in its gentle approach. But then, everything began to go awry. Progress was still moving forward, but it had moved out of the cosmonauts' field of view. They dashed from one porthole to another to find its hiding place within the dark sky. But when they finally glimpsed it again, it was too late. Progress was too close, and moving too quickly. In desperation, Tsibliyev struggled to regain control of the vehicle with the help of the two hand levers. Progress began a long course around Mir until, with horror, Tsibliyev saw the image of the space station, his space station, advancing relentlessly towards him on the screen. Mir quivered as Progress struck the *Spektr* module. An alarm immediately went off, warning the three cosmonauts that pressure was falling in the station. The impact had caused a tiny rupture in the walls of *Spektr* and its air was leaking out. Either they would have to evacuate Mir or isolate the leaking module by closing the airlock that led into it. Bravely, the Russians chose the second solution and their American colleague fully supported them, at great risk to their own lives. Inside the station, the pressure was normally maintained at 780 milli-metres of mercury, but the atmosphere was rapidly rarefying. Within a few minutes, the pressure had gone down to 670 millimetres, as though the three men had suddenly found themselves at the top of a 4000 metre-high mountain. In a few more minutes, they would very likely lose consciousness and death would follow shortly afterwards. Following a desperate struggle with

■ The *Discovery* cockpit.
Aboard the space shuttle,
astronauts have the
impression of being inside a
planetarium. Broad
windows look out in all
directions, with
spectacular views of
fellow astronauts at
work in the cargo bay,
and magnificent
Earth beyond.

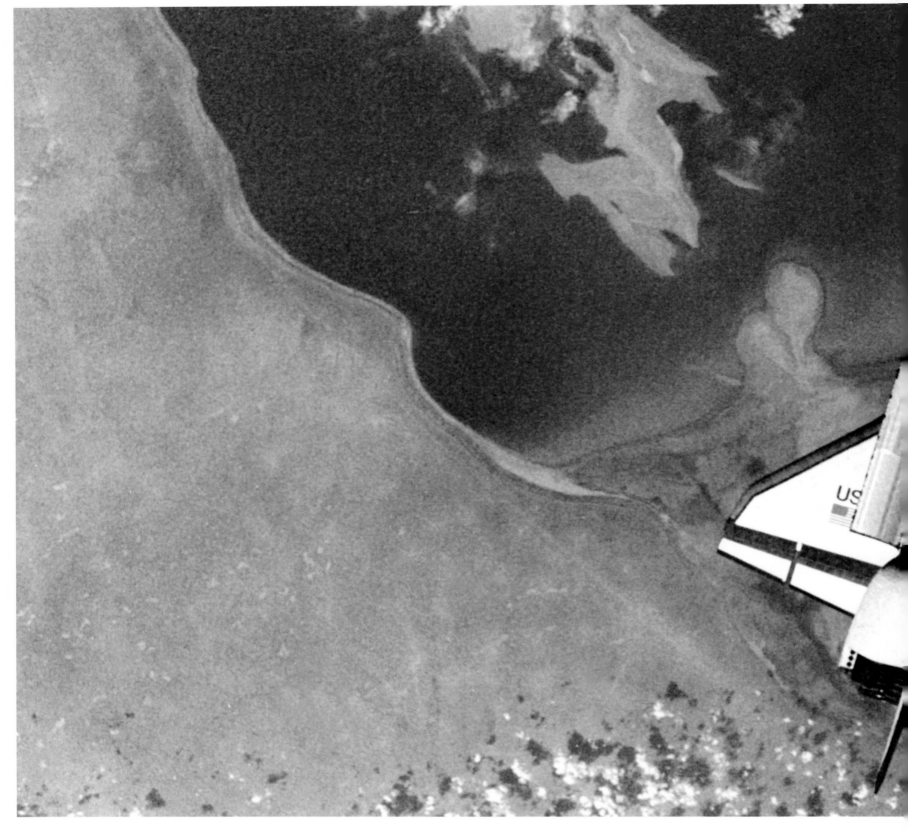

the hatchway into the *Spektr* module, which the pressure of outpouring air prevented them from closing, Michael Foale and Aleksandr Lazutkin finally devised their own solution. They placed a plaque across the airlock and the pressure difference held it there without further intervention. At least in the short term, the station was saved, and its three occupants with it. By further ingenious measures, the three men would still have to restore the damaged electricity system. Solar panels had been destroyed along with the *Spektr* module. In addition, the impact had caused Mir to spin slowly and this meant that the remaining solar panels were no longer correctly deployed in relation to the Sun. Finally, it remained to cut off the damaged module with a special airlock built on Earth and carried up to the station by another Progress vessel.

Following this rescue operation, astronauts and cosmonauts resumed missions aboard the station, although the *Spektr* module was never used again, remaining as empty and cold as the void surrounding it. Mir was scheduled for permanent closure on 28 August 1999, at the departure of the crew comprising Viktor Afanasyev, Sergei Avdeyev and Jean-Pierre Haigneré. By then it had already witnessed ten years of uninterrupted occupation by about thirty crews. However, the Russian space agency had been seeking foreign partners ready to

pay to keep it in orbit, at a cost of around 80 million euros per year. The hope was that on 21 January 2001, the Mir crew would be able to celebrate the fifteenth anniversary of the station's arrival in space. Will history recall the intervention of Mircorp, a private company that took Mir's destiny into its hands, offering it as a research laboratory, but also as an advertising studio, television studio and even as a spaceborne hotel for billionaires? On 6 April 2000, a new crew made up of Sergei Zalyotin and Aleksandr Kaleri arrived on the station to tidy it up in readiness for any wealthy clients that might soon turn up, guaranteeing its existence until the following year. Mir undoubtedly deserved better than this. Like Salyut 7 before it, Mir could well have carried the flame to the next generation, the International Space Station. Mir was the pride of the Russian people and they could not resolve themselves to abandon this symbol of Soviet space conquest. And yet the majestic cosmic caravel could not remain in orbit forever. Soon it would have to end its days in the manner of any other old spacecraft, as a glorious shooting star burning up in the fire of the Earth's atmosphere.

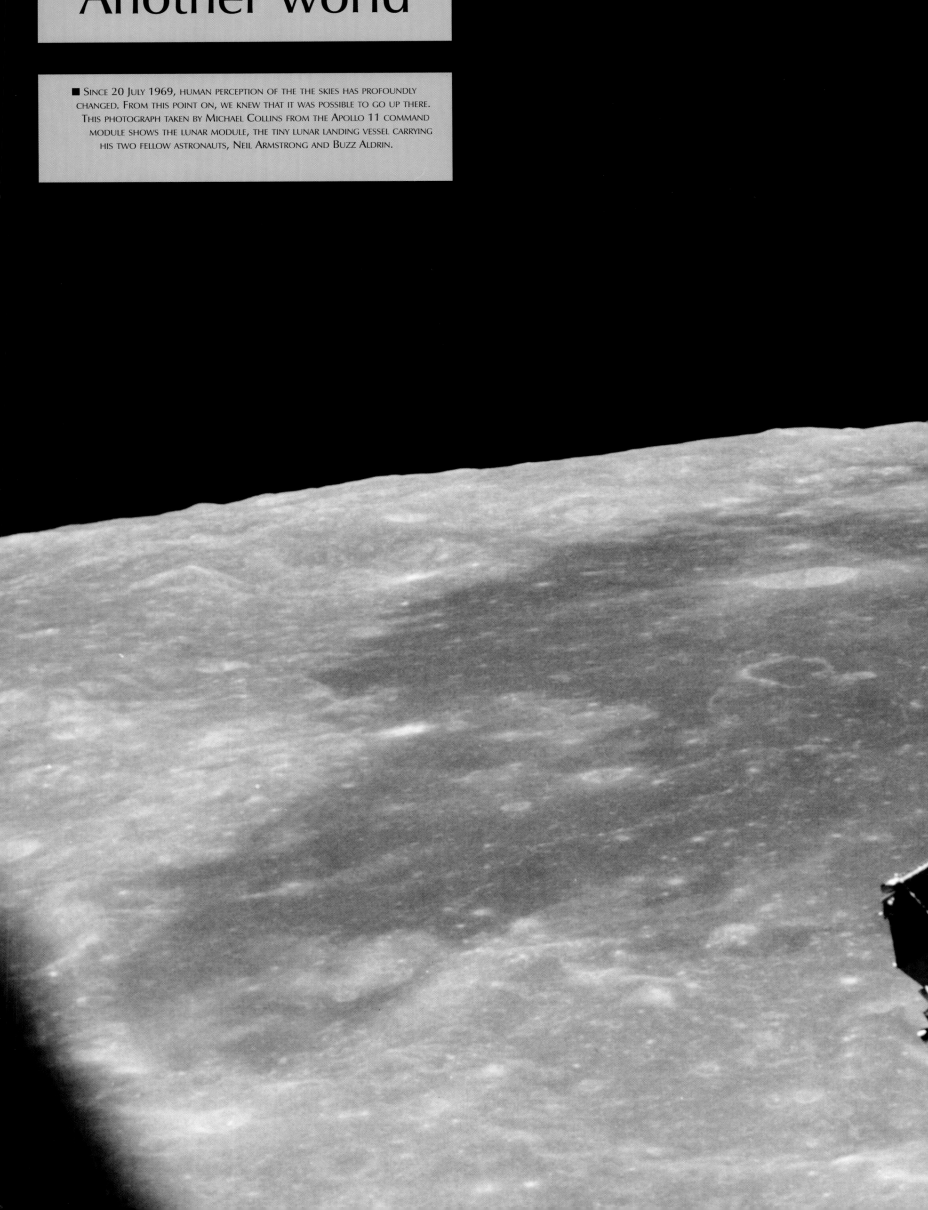

Another world

■ Since 20 July 1969, human perception of the the skies has profoundly changed. From this point on, we knew that it was possible to go up there. This photograph taken by Michael Collins from the Apollo 11 command module shows the lunar module, the tiny lunar landing vessel carrying his two fellow astronauts, Neil Armstrong and Buzz Aldrin.

■ SOMEWHERE IN THE SEA OF TRANQUILLITY, PHOTOGRAPHER NEIL
ARMSTRONG AND HIS SUBJECT BUZZ ALDRIN ANSWER ONE ANOTHER
ACROSS THE CHASM OF SPACE THROUGH THE PLAY OF LIGHT IN THE
GOLD OF AN ASTRONAUT'S HELMET.

The hatch opened noiselessly. Ronald Evans, pilot of the Apollo 17 command module, seemed to float inside his space suit. Although it weighed nothing, the space suit still possessed a considerable mass and behaved with the corresponding inertia. It seemed to have a life of its own. When the astronaut made a simple movement, his feet left their soles behind them whilst his forehead knocked gently against the thick visor of the helmet. However, the clumsy outfit was not really a handicap and, having spent the past week circling the Moon alone in his module, Ronald Evans was only too glad to stretch his legs for a while. His two companions, Gene Cernan and Harrison Schmitt, had come back two days before and since then the three men had been living in the cramped confines of the command module. Schmitt and Cernan had just returned from the Moon where they had spent three days exploring with the help of their electric jeep and enjoying the wonderful scenery of the Taurus–Littrow landing site. The space walk that Ronald Evans was about to engage upon was thus a kind of reward for the man who had taken on the most frustrating role of the Apollo 17 mission, patiently watching over his fellow astronauts from his lunar orbit.

For precisely one hour and six minutes, Ronald Evans would enjoy an extraordinary experience that had only been shared by two other astronauts in the history of space flight: Alfred Worden and Thomas Mattingly. Far from Earth, clinging to his frail skiff, he would drift across the waves of space. Such an excursion into deep space is an inexpressible experience that the Apollo astronauts have to a large extent and with a certain modesty kept to themselves. It was 17 December 1972. Evans perched trembling on the Apollo 17 command and service module, his heart in his mouth, completely lost in the middle of nowhere, an infinitesimal satellite of the Solar System, falling into the vertiginous abyss. In the ink black sky that entirely surrounded him, he turned his back on the Sun's intolerable glare and saw only the enormous, dazzling face of the Moon, almost full. Evans was halfway between Earth and Moon, a tiny figure standing on a nutshell that would soon take him back home. In order to pick out his own world, he had to lower his helmet with its protective golden sheen and eclipse the Sun with a large gloved hand. There it was in the shadows, a gentle crescent of pale bluish light, so small and distant across the immense black sea. Lost in the void, speeding homeward at 10 km/s, Ronald Evans had the strange impression of being perfectly motionless. The scene was quite unreal. Over to one side, the larger-than-life lunar lantern with its fixed glow seemed to have been overdone. The image was too clear, too steady. Evans could name all the craters

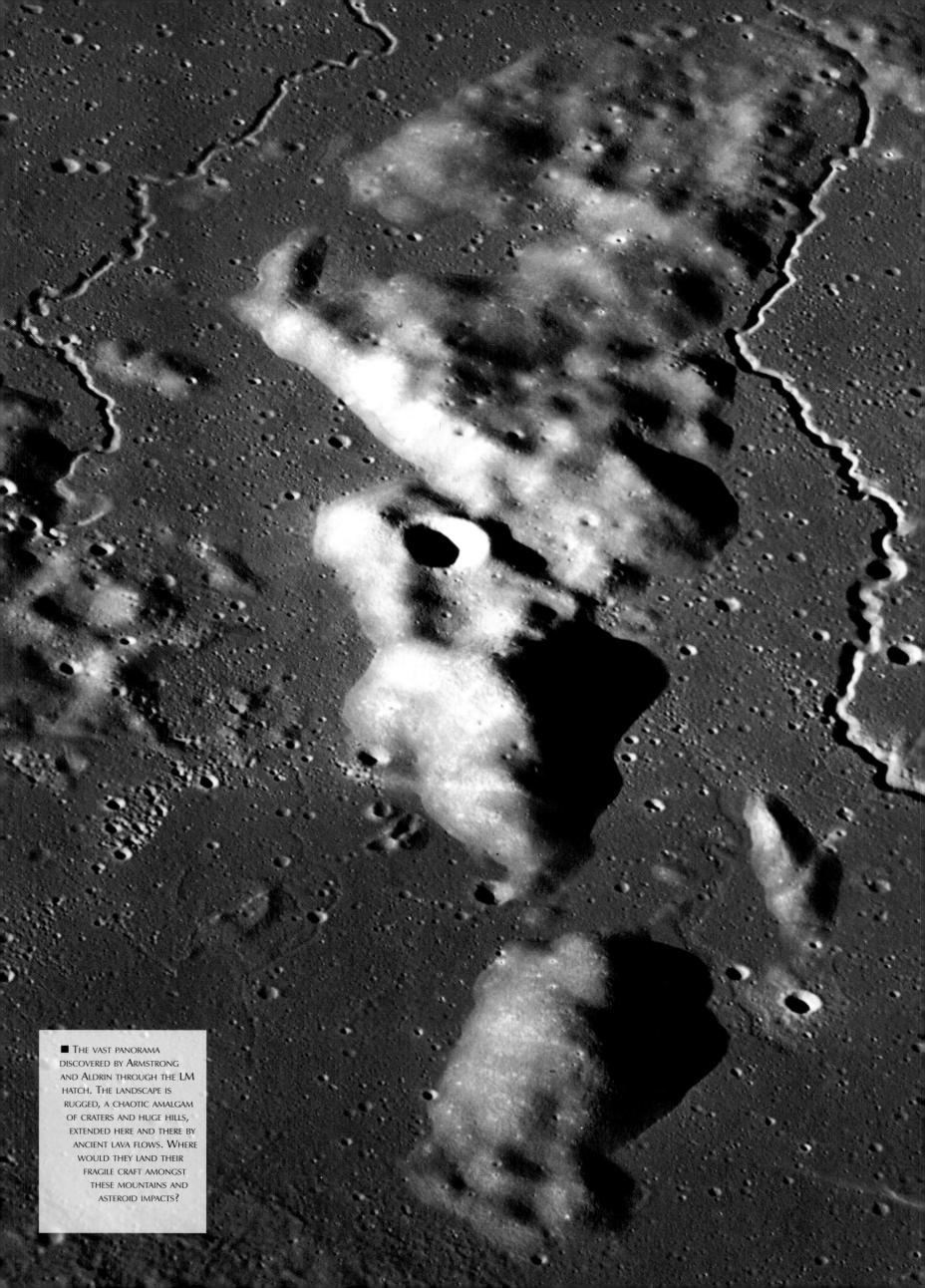

■ THE VAST PANORAMA
DISCOVERED BY ARMSTRONG
AND ALDRIN THROUGH THE LM
HATCH. THE LANDSCAPE IS
RUGGED, A CHAOTIC AMALGAM
OF CRATERS AND HUGE HILLS,
EXTENDED HERE AND THERE BY
ANCIENT LAVA FLOWS. WHERE
WOULD THEY LAND THEIR
FRAGILE CRAFT AMONGST
THESE MOUNTAINS AND
ASTEROID IMPACTS?

visible to the naked eye, projecting their stark shadows along the terminator: Copernicus, Kepler, Aristarchus, Gassendi and many more. It looked like a stage set or a crazy demiurgic planetarium. Moving slowly along the outer face of the module, Evans recovered the canisters containing the photographs he had so methodically taken during his week in lunar orbit. When he went back inside the spartan but hospitable module *America* and closed the airlock behind him, a shiver ran up his spine. Overwhelmed, he suddenly realised that he had lived through a most uncommon experience.

With the return journey to Earth, Ronald Evans, Gene Cernan and Harrison Schmitt brought off one of the most extraordinary human adventures of all time. For two more days yet, the three men would share this unbelievable space odyssey, exchanging stories of a Moon walk, exploration in the lunar rover, or an unparalleled walk in space. Satisfied, but at the same time nostalgic, the three men were already wondering when the next crew would be leaving for the Moon. However, Apollo 17 was to be the last of NASA's manned lunar missions.

LANDING A MAN ON THE MOON: THE GREAT CHALLENGE

Everything began fourteen years before Apollo 17. Cold War America, ruffled by Soviet success in the cosmos, began to fear that their archenemy might be the first to gain control of circumterrestrial space. In particular, the concomitant growth of military technology in this new environment seemed to pose a real threat. The US therefore threw itself headlong into the technological and ideological chase commonly referred to as the space race, hoping if possible to make up for lost time and even to take the lead. Indeed, the Soviets had already achieved a long series of firsts since the beginning of the space race. They were first to get a satellite into orbit, with Sputnik on 4 October 1957. They were first to send a living creature into space, when they dispatched the dog Laika on 3 November 1957. They were the first again when they sent space probes Luna 1, 2 and 3 around and onto the Moon between January and October 1959. But all this was just a prologue to the crowning achievement: they were the first to send a man into space, when Yuri Gagarin orbited the Earth on 12 April 1961.

Stupefied at first by this series

of Soviet successes, the Americans gradually began to put together an organisation capable of competing with Sergei Korolev's inspired and ambitious team. This was the National Aeronautics and Space Administration, usually referred to as NASA, created on 1 October 1958. By 5 January 1959, when the new organisation unveiled its ambition to send a man into space, NASA already employed several tens of thousands of people. The Mercury programme was launched with seven aspiring astronauts soon to become living legends: Gordon Cooper, John Glenn, Virgil Grissom, Walter Schirra, Alan Shepard and Donald Slayton.

On 25 May 1961, just a little more than a month after Yuri Gagarin's historic flight, the freshly elected president of the United States, John Fitzgerald Kennedy, delivered his state of the union address before the US Congress. He made the following declaration that would change the course of twentieth-century history: 'I believe that this nation should commit itself to achieving the goal, before this decade is out, of landing a man on the Moon and returning him safely to the Earth. No single space project will be more exciting, or more impressive to mankind, or more important, and none will be so difficult or expensive to accomplish.' Kennedy threw out this unlikely challenge to the Russians right in the middle of the Cold War. A month before the disastrous Bay of Pigs operation had ended in total failure, and the Eastern Block would soon erect the Berlin Wall. Eighteen months later, the psychological war between the two opposing regions culminated in the Cuban missile crisis. For a few years, the Russians added to their list of successes in space, but the formidable American economic and technological machine had been set in motion and nothing could now stop it.

Although at first they seemed reluctant to accept the challenge, the Soviets, in the greatest secrecy, had quickly set up their own space programme with the Moon as its target. It was no coincidence that, at the outset, the two nations were technologically very close. Indeed, they had chosen to follow the same path, inspired by Wernher von Braun's notorious V2 missiles. These weapons, and the engineers who developed them for Nazi Germany, were soon recovered by the United States and the USSR at the end of World War II. Moreover, in the middle of the Cold War, satellite launch vehicles were developed in parallel with intercontinental ballistic missiles designed to carry nuclear

■ SEEN FROM THE SLOWLY DESCENDING LM, THE LUNAR SURFACE REVEALS AN ASTONISHING FRACTAL STRUCTURE. ON EVERY SCALE, THE SAME LANDSCAPE APPEARS. IN EVERY DIRECTION, THERE ARE CRATERS, HILLS AND ROCKS OF SPECTACULAR DIMENSIONS.

■ SINCE THE BIRTH OF THE SOLAR
SYSTEM AROUND 4.5 BILLION YEARS AGO,
THE MOON HAS BEEN IMPACTED BY
LITERALLY MILLIONS OF GIANT METEORITES
AND EVEN ASTEROIDS MEASURING
SEVERAL KILOMETRES ACROSS. MOST OF
THE IMPACT SITES VISIBLE THROUGH THE
TELESCOPES OF AMATEUR ASTRONOMERS
ON EARTH DATE FROM ONE TO FOUR
BILLION YEARS AGO. FROM THEIR
PRIVILEGED VIEWPOINT IN LOW LUNAR
ORBIT, APOLLO ASTRONAUTS WERE
STRUCK BY THE BEAUTY OF THE
LARGER CRATERS, THEIR RIMS
SCULPTED INTO IMPRESSIVE
TIERS OF TERRACES.

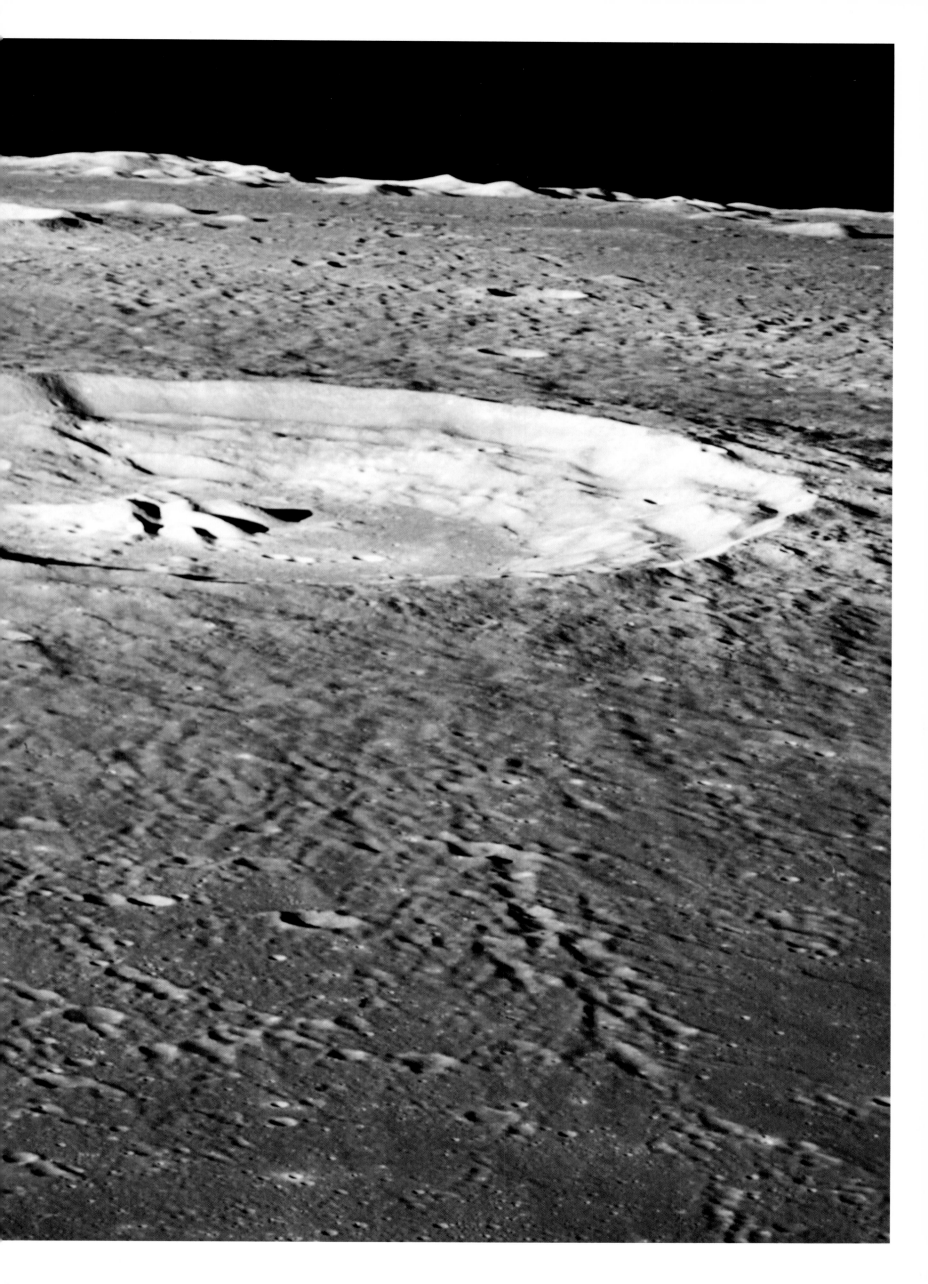

warheads. In fact, exactly the same masses and flight techniques are involved. At the beginning of the 1960s, the United States lagged behind and could only sit back and admire the impressive series of Soviet achievements, a list that was soon to be continued. On 6 August 1961, Gherman Titov smashed the record time spent in space. He returned safe and sound after an astonishing twenty-four hours in orbit, having spun round the Earth seventeen times. On 16 June 1963, Valentina Tereshkova became the first female cosmonaut, aboard Vostok 6. Then on 12 October 1964, Konstantin Feoktistov, Boris Yegorov and Vladimir Komarov made up the first multi-man crew aboard Voskhod 1. Finally, on 18 March 1965, Aleksei Leonov became the first man to leave the protective enclosure of his spacecraft and venture into empty space.

THE WIND CHANGES IN THE RACE TO THE MOON

This was the last victory for Soviet astronautics. Slowly but surely the United States was making headway and revenge was in the air. Once the Mercury programme had been concluded, hastily put together for stuff-of-heroes astronauts, NASA turned its attention to more serious business. With the Gemini programme, they aimed to confirm technical and strategic choices for their future conquest of the Moon. It marked the beginning of a spectacular series of American successes. In August 1965, Gordon Cooper and Charles Conrad finally secured a record for the United States, remaining in Earth orbit for a whole week inside their Gemini 5 capsule. Very soon, Frank Borman and James Lovell had taken the record to almost two weeks aboard Gemini 7. For the first time, during the same Gemini 7 flight, two spacecraft flew in symphony, when Borman and Lovell were joined in orbit by Walter Schirra and Thomas Stafford. Aboard the Gemini 6 capsule, the latter managed to come within one metre of their Gemini 7 colleagues. It remained only to place the keystone of the lunar programme, a link-up in space. This was soon achieved when Gemini 8 docked with the Agena spacecraft in July 1966.

But the two great world powers, now engaged in a frantic race, were starting to move too quickly. On 27 January 1967, Virgil Grissom, Edward White and Roger Chaffee climbed into their Apollo 1 cabin for a routine

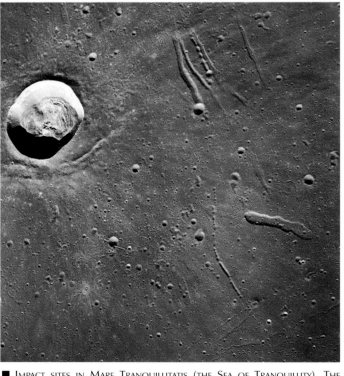

■ IMPACT SITES IN MARE TRANQUILLITATIS (THE SEA OF TRANQUILLITY). THE MOON IS COVERED WITH MILLIONS OF BILLIONS OF CRATERS OF ALL SIZES. THE SMALLEST ARE MICROSCOPIC, WHILST LARGER ONES LIKE MARE IMBRIUM (THE SEA OF RAINS) MEASURE HUNDREDS OF KILOMETRES IN DIAMETER.

training session. A short-circuit threw out a spark which set alight the pure oxygen atmosphere in the cabin, causing a violent conflagration within a few seconds. The three astronauts were asphyxiated. On the Soviet side, as on the American, many unmanned launch vehicles had already been lost at blast-off. However, the first fatal accident came only three months after the Apollo 1 tragedy, when Vladimir Komarov was killed aboard his new Soyuz 1 capsule on 23 April 1967. The blow was hard to bear, on both sides of the Iron Curtain. At Baikonur, the engineers and technicians were all the more at a loss because the legendary Sergei Korolev, father of the Soviet space programme, had died only the previous year. To worsen the situation, politicians and engineers were divided over the strategy they should adopt for lunar conquest. With forty-four engines to propel it, the monstrous N1 rocket was designed to take humans to the Moon. It weighed 2700 tonnes and measured 112 metres tall. However, it exploded in flight after the first four launches and the USSR subsequently abandoned the idea of walking on the Moon.

In the United States the shock of the terrible accident proved to be salutary. The Apollo programme was stopped for almost two years so that it could be completely reworked. Launch vehicles and modules were modified and improved, with the emphasis on reliability, and on 11 October 1968, Apollo 7 left for space with Walter Schirra, Walter Cunningham and Donn Eisele on board. The Moon rocket, the little space train it carried, was tested for the first time in Earth orbit and everything went according to plan. The great Apollo story had begun. On 21 December 1968, Frank Borman, James Lovell and William Anders left for the Moon aboard Apollo 8. For the first time in history, men left their planet and went out to discover a new celestial body. On the evening of Christmas day, the Apollo 8 crew went into orbit around the Moon, from which vantage point they were able to see its hidden face. Orbits were carefully checked and all equipment tested. With growing confidence, NASA scheduled two further launches to qualify the Apollo programme for the great day. In March 1969, David Scott, James McDivitt and Russel Schweikart flew Apollo 9 which comprised all the components of the future flight to the Moon: command module, service module and lunar module (also known for a time as the Lunar Excursion Module, the famous LEM). It only remained for

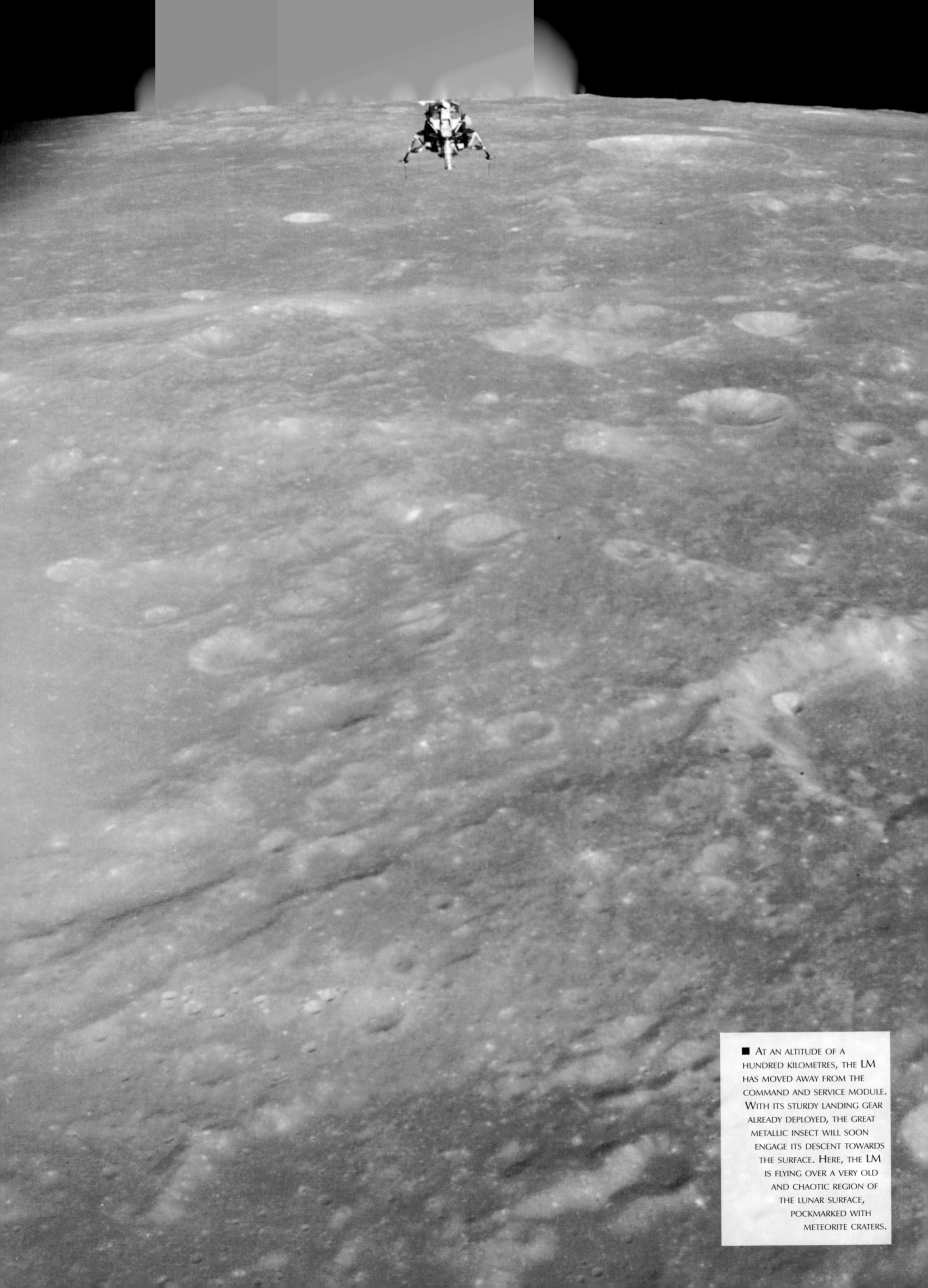

■ At an altitude of a hundred kilometres, the LM has moved away from the command and service module. With its sturdy landing gear already deployed, the great metallic insect will soon engage its descent towards the surface. Here, the LM is flying over a very old and chaotic region of the lunar surface, pockmarked with meteorite craters.

John Young and Gene Cernan, who did not yet know that they would one day walk on the Moon, and Thomas Stafford, who would not share this honour, to carry out what was surely the most frustrating mission in the whole history of space conquest. For it fell upon them to play out the final rehearsal, a real-life simulation of lunar landing. Apollo 10 flew from Cape Canaveral on 18 May 1969. The flight went exactly as planned. Once in lunar orbit, the LM was detached from the two Apollo craft that made up the little space train. Piloted by Stafford and Cernan, it made a slow descent towards the surface, stopping only fourteen kilometres from the target coveted by a whole nation for eight years. The two astronauts had to demonstrate uncommon self-control not to actually land the lunar module. They already knew that this extraordinary destiny had fallen upon the lucky crew that would succeed them, aboard Apollo 11.

COUNTDOWN

On the morning of 16 July 1969, the weather was clear at Cape Canaveral. On the launch pad, the giant Saturn V rocket was ready for lift-off. Right at the top, settled in their cramped command module for two hours now, Neil Armstrong, Buzz Aldrin and Michael Collins awaited deliverance. They were thinking about the genuine war effort their country had instigated to reach the Moon before the Soviet Union. At its foundation in 1958, NASA took on about 10 000 people. By 1969, almost 40 000 were employed directly by NASA, whilst 350 000 other technicians and engineers working for the aeronautic industry were in some way involved in the Apollo programme. Apart from these, hundreds of thousands of specialists in new materials and the nascent area of computer science should be added to the list. Indeed, such domains owe a large part of their future explosion to this same Apollo programme. In his book *A la*

■ THERE ARE NO PHOTOGRAPHS OF NEIL ARMSTRONG ON THE MOON. HOWEVER, DURING THE WHOLE OF THE FIRST MOON WALK, WHICH ONLY LASTED TWO HOURS OR SO, ARMSTRONG TOOK MANY PHOTOGRAPHS OF HIS COLLEAGUE, STRANGE EXPLORER OF ANOTHER WORLD.

conquête de la Lune, Jacques Villain suggests that as many as ten million people across the world contributed in some way to the American conquest of the Moon. Along with the construction of the Great Wall of China, the Manhattan project and certain exorbitantly expensive military programmes, Apollo is probably one of the most costly human undertakings ever realised. It is estimated that a total of 100 billion euros were spent on achieving this dream. The colossal price paid for the incredible technological and ideological competition between the world's two most powerful nations, combined with the astronomical cost of the arms race, eventually brought about the economic ruin of the Soviet empire. Will history recall that, at the end of the second millennium, one of humanity's most thrilling adventures was actually the fruit of a hard fought ideological war rather than a quest for knowledge or even a simple desire to explore our cosmic environment?

On the 16 July 1969, Armstrong, Aldrin and Collins already knew that the Soviets had lost the race to the Moon. But what was lacking in order to turn the USSR's initial successes into lunar conquest, to bring their early triumphs in space to an apotheosis? On one side, bureaucrats obsessed with secrecy kept a heavy hand on Korolev's sure and efficient development, mainly via the military machine, imposing inappropriate constraints on such an immensely difficult task as a journey to the Moon. On the other, the engineers of a great liberal economy worked together in relatively transparent conditions where technological innovation was the only framework for thought. To worsen the situation, there were dissensions within the USSR, with three competing Moon programmes (put forward by Korolev, Yangel and Tchelomei), later whittled down to two, using the N1 and the UR 500 Proton. Added to this, the Soviets failed to make two crucial technological leaps that might have got their gigantic N1 rocket safely off the ground. Like the faithful Semiorka,

■ Despite the gentle shape of its mountains and the neutral tones of its dusty seas, the Moon is an unbelievably hostile environment. The day lasts fifteen terrestrial days during which a blinding Sun bombards the surface with ultraviolet, X and gamma radiation, not to mention a deluge of high energy particles. When night falls, this torrent of deadly particles peters out for two weeks. However, the star-filled sky remains a lethal threat to astronauts under an endless storm of cosmic rays from distant stellar explosions.

■ JAMES IRWIN CHECKS THE LUNAR ROVER AND PREPARES TO DISCOVER THE WONDERFUL LANDSCAPE OF THE LUNAR APENNINES. BEHIND THE APOLLO 15 ASTRONAUT LOOMS THE AUSTERE AND MAJESTIC SILHOUETTE OF MOUNT HADLEY, 5000 METRES HIGH.

the N1 was propelled by reliable but relatively small engines. Each could only muster 150 tonnes of thrust, so that no fewer than forty-four were needed to get the rocket into orbit. For their part, the Americans had opted for engines of quite impressive dimensions, each producing a thrust of 700 tonnes. This meant that the American rocket required far fewer engines, only five for the first stage, compared with thirty for the N1. Moreover, the second and third stage engines of the American rocket ran on a far more efficient propellant than the kerosene-liquid oxygen mixture used by the Soviets. American engineers had opted for a liquid hydrogen-liquid oxygen mixture. Finally, it may be said that the Americans virtually put together modern computer science from scratch for the purposes of the Apollo mission, whilst their Soviet counterparts were unable to develop microcomputers that could control the flight of a giant rocket.

AN EARTH-SHATTERING ROAR

At 9.30 a.m on 16 July 1969, Cape time, the most powerful rocket ever built prepared to leave Earth. The million or so people who had gathered enthusiastically and in indescribable chaos around the Cape Canaveral site, along the Atlantic beaches, and around the small town of Cocoa Beach had difficulty concealing their emotions. Most were too far away to glimpse Saturn V on launch pad 39A. The spacecraft stood 110 metres tall and measured more than 10 metres in diameter at the base. Altogether it weighed 2950 tonnes and its huge tanks were filled with 2760 tonnes of liquid propellant.

The three men lay side-by-side in the tiny command module that would eventually return them to Earth after their epic journey. Above them was the launch escape tower that would expel the module violently upwards in the event of any incident at blast-off. Beneath the command module was the service module. Equipped

with a powerful engine, it would eventually bring the crew back to Earth orbit. The combined command and service module (CSM) would be their home during the journey and would await them in lunar orbit during the lunar landing. Finally, below the service module, with the four legs of its landing gear folded beneath it to save space, was the lunar module *Eagle*. Connected upside-down underneath the engine nozzle of the service module, once again for reasons of space, it would have to be detached, turned around and redocked in the right direction by the crew during the translunar crossing. The whole space train weighed more than 40 tonnes.

At 9.32 a.m. the five main engines of the first stage were ignited, developing a combined thrust of almost 3500 tonnes and burning about 15 tonnes of fuel per second. With disconcerting and dreamlike lethargy, the Saturn V rocket uttered an Earth-shattering roar and began to rise towards the sky. The journey from the Earth to the Moon that Jules Verne had dreamed of more than a century before, and which by some strange intuition he had also imagined taking off from the point of Florida, had finally begun.

The blast-off of a Saturn V rocket is an incredible experience, not only for the three men shaken around inside the command module and boosted upwards by 150 million horsepower of thrust, but also for the incredulous public, tears in their eyes, watching the fragile white silhouette lifting off, trailing behind it a blinding flame half a kilometre long. After climbing for about twelve minutes, Armstrong, Aldrin and Collins were already in Earth orbit, ready for their great leap towards the Moon. Less than three hours later, having completed a magnificent world tour at 28 000 km/hr, Armstrong switched on the single third stage engine. A staggering thrust, hard to bear after any length of time spent in weightlessness, boosted them suddenly up to 39 600 km/hr. This is

the Earth escape velocity, required to tear the spacecraft completely away from Earth's gravitational bonds. Once the third stage of the rocket had completely burnt out, the space train jettisoned its cumbersome companion and began its journey unburdened across the void of space. But this journey would not last long. Despite appearances, the Moon is not situated at an astronomical distance from us! 384 000 kilometres is the distance travelled by a commercial jet every month, or a London or New York taxi in under five years. In the context of space travel, Mir covered the same distance around the Earth in about twelve hours, and in cosmic terms, light takes only a little more than one second to accomplish this tiny step.

The Apollo 11 crew would take only three days to make the Earth-Moon crossing. During the journey, they detached the command and service module *Columbia* from the lunar module *Eagle* as planned, turning the latter around and redocking it onto the command module. On the afternoon of 17 July, the three men flew over the Moon at a mere 100 kilometres altitude. The time had come for Michael Collins to part company with his fellow travellers, in an atmosphere charged with emotion. Neil Armstrong and Buzz Aldrin closed the LM airlock behind them, leaving Collins alone in the command and service module. The LM then undocked from the command module and moved several metres away, under the scrutiny of Michael Collins who checked that the four outstretched legs of the LM landing gear were correctly deployed. Slowly, *Eagle* and *Columbia* moved apart and the LM gently engaged its descent towards the lunar surface. Neil Armstrong and Buzz Aldrin were wearing their space suits in readiness. They were standing upright, side by side, firmly strapped into safety harnesses. Between them was the LM control panel, and opposite were two triangular windows through which they could watch the 'magnificent desolation' of this other world

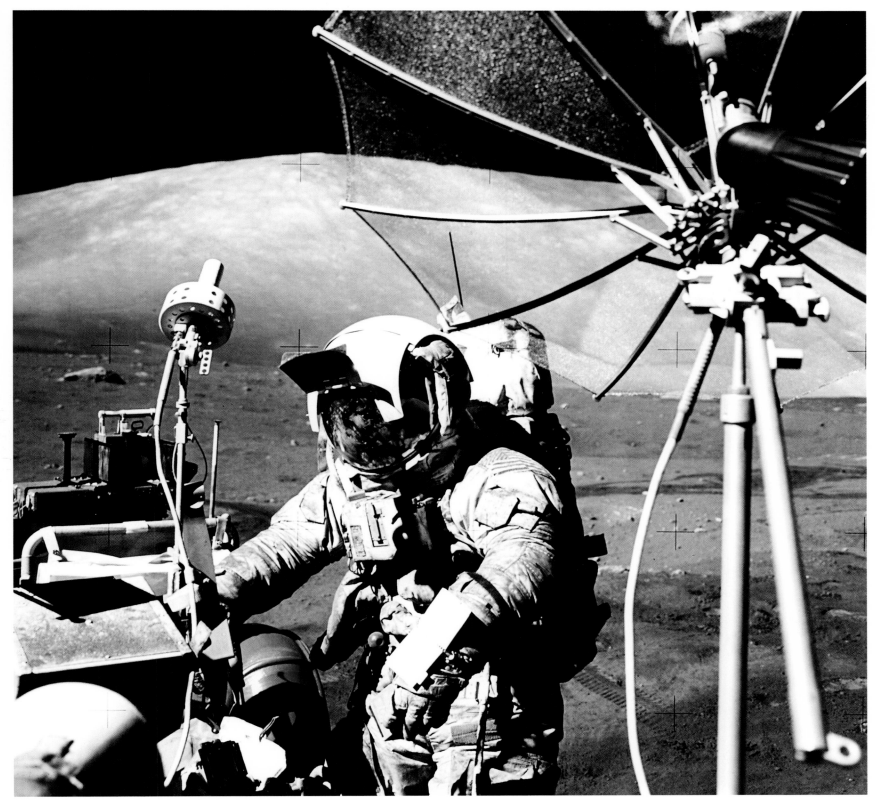

approaching, to use the term later employed by Buzz Aldrin. Neil Armstrong was at the controls, whilst Buzz Aldrin checked the descent path

apart from a difference of scale, and low hills literally everywhere, also repeating the same shapes *ad infinitum*. This would have made it

parameters. In fact, at this point the flight was in automatic mode, entirely controlled by the onboard computer. When they arrived at an altitude of about 16 000 metres, with another 300 kilometres to go in order to reach the landing site chosen in the gentle plain of the Sea of Tranquillity, the computer fired the LM descent engine. The lunar module abruptly slowed down and yielded itself up to the Moon's field of gravity, plunging down towards the surface. Seen from within, the sight would have been totally disconcerting if the two astronauts had not spent hundreds of hours rehearsing their mission, or if they had not known the area where they would soon be landing to the last detail. The fractal nature of the lunar landscape would have been particularly disorientating, pockmarked by hundreds of craters of all sizes, but all identical

quite impossible to estimate the distance to the surface, without the altimeter under the firm scrutiny of Buzz Aldrin. They arrived at 1800 m after eight minutes of descent and the two men were beginning to feel that their mission was on the point of succeeding, when suddenly the alarm began to sound and Buzz Aldrin saw a yellow caution light flashing insistently on the computer. Whilst Aldrin stood with his finger poised on the abort button that would fire the LM ascent engine and trigger an emergency return to lunar orbit, Armstrong used the craft's intercom to ask the Houston control centre what he should do. The reply seemed a long time in coming, firstly because the technicians were looking for the problem that had been signalled and were trying to gauge its gravity, and secondly because a

conversation from the Moon to Earth and back again involves a time lag of three full seconds. At last, NASA engineers gave the green light to the astronauts to continue the descent. The program alarm simply indicated that the onboard computer had reached executive overflow and was struggling to process the thousands of operations per second required to control the flight: adjusting the descent engine thrust, checking the attitude and altitude of the LM, now falling vertically, and so on. A reminder that this was in 1969, before the extraordinary development of microcomputers. A second and then a third alarm went off, at 1000 and 600 metres altitude, respectively. Each time NASA engineers gave the go-ahead to Armstrong and Aldrin, confident in the artificial intelligence of the lunar spacecraft. Meanwhile, the two astronauts could concentrate on landing preparations. Four hundred metres. The descent engine had been burning for about ten minutes. Aldrin realised that a large part of their stock of propellant had been used up, since the flight had lasted longer than expected. Much more seriously, the two American pilots could see through their portholes that they had missed the scheduled landing site by several kilometres. The automatic pilot aboard the LM was taking them straight towards a large crater, filled with boulders and rocks of all shapes and sizes. It would be quite impossible to land the LM on such a site. If the craft fell over or was damaged upon touchdown, the first two men to land on the Moon would also be the first two to die there. Back on Earth, Houston flight controllers recorded a sudden increase in Armstrong's heart rate, to almost 160 beats per minute. The pilot had taken manual control of the

■ IF THERE IS A SINGLE TWENTIETH-CENTURY ASTRONAUT WHO SHOULD GO DOWN IN HISTORY, IT IS JOHN YOUNG. HE HAS BEEN INTO SPACE SIX TIMES AND TRAVELLED TWICE TOWARDS THE MOON. HE IS A SPACE SHUTTLE PILOT AND HAS ALSO DRIVEN A

LM and, by a slight thrust of the engine, was steering his vehicle downrange of the chaotic landscape, seeking out a flat area to touch down. With only 25 metres to go, Aldrin could now forget the abort button. The LM was too low to recover in an emergency. There was no option but to land and time was running out. A new alarm rang out on board, to inform them that the descent engine tanks had almost run dry. Imperturbable, Armstrong continued the delicate descent, whilst Aldrin looked fixedly at the altimeter, so as not to see the flashing caution light on the propellant gauge. Twenty metres, fifteen metres, ten metres. On the Moon there is not the slightest trace of an atmosphere. If the engine were to stop, the LM would simply crash into the surface. But then the miracle happened. Through their portholes, the two men could make out dust swirling around them as the flame of the engine stirred up the lunar dusts. A pitch black shadow appeared, moving slowly across the ground, the phantasmagoric silhouette of a gigantic iron beetle spewing fire. Then the four green lights corresponding to the four landing gears lit up one by one and at last Neil Armstrong switched off the engine of his spacecraft and uttered the unforgettable words: 'Houston, Tranquillity Base here. The Eagle has landed!'

20 JULY 1969, 20H 17M

The time was 20h 17m GMT on 20 July. Apollo 11 had touched down just at the edge of the Sea of Tranquillity in an attractive site

much favoured by Earth-based amateur astronomers. Through their telescopes they could admire the sunrise at the Sabine and Ritter craters, and nearby, discover the gentle shadows cast by the motionless wrinkles and strange domes of the Sea of Tranquillity. Beyond were the narrow, parallel grooves of Moltke, known as rilles, bordering a much older terrain where the intrepid *Eagle* would have been in great difficulty. Looking through the porthole, Armstrong described the landscape: 'The area out the left-hand window is a relatively level plain cratered with a fairly large number of craters of the five to fifty foot variety; and some ridges (which are) small, 20, 30 feet high, I would guess; and literally thousands of little, one and two foot craters around the area. We see some angular blocks out several hundred feet in front of us that are probably two feet in size and have angular edges. There is a hill in view, just about on the ground track ahead of us. Difficult to estimate, but might be a half a mile or a mile… I'd say the colour of the local surface is very comparable to that we observed from orbit at this Sun angle – about 10 degrees Sun angle, or that nature. It's pretty much without colour. It's gray; and it's a very white, chalky gray, as you look into the zero-phase line. And it's considerably darker gray, more like ashen gray as you look out 90 degrees to the Sun. Some of the surface rocks in close here that have been fractured or disturbed by the rocket engine plume are coated with this light gray on the outside; but where they've been broken, they display a very dark gray interior; and it looks like it could be country

basalt.' Six and a half hours later, having recovered from the excitement of his magnificent landing – he learnt afterwards that precisely sixteen seconds of fuel was left at the moment when he had cut the engine – Neil Armstrong was ready to set foot on the Moon. He came out onto the upper step of the stairway, with his gloved hand firmly grasping the thin rail, and looked across the grey variations of the surrounding landscape. Then, one by one, he descended the eight rungs of the ladder that ran down one leg of the lunar module.

On the last step, he hesitated for a brief instant, then lowered the thick boot of his pressurised space suit onto the ground of this other world, before declaring for posterity: 'That's one small step for a man, one giant leap for mankind.' Shortly afterwards, Buzz Aldrin joined his companion and they were able to discover together the new sensations offered by another planet. Gravity is six times weaker on the Moon than on Earth. With their cumbersome space suits, the two men would each have weighed about 180 kilos on Earth, whereas here, they weighed less than thirty kilos. Although they felt light, their movements were inhibited for other reasons: firstly, because of the air pressure inside the suits, and secondly because inertial mass remains the same even when weight decreases. Thus the slightest push with the foot would lift them off the ground, but then it was difficult to stop the motion or to twist around. It was as though some invisible hand were leading them forward. They soon gave up trying to walk and adopted a kind of swaying trot, interspersed with slightly longer jumps. Everything seemed so easy now. Neil Armstrong continued:

■ Captured by Neil Armstrong's camera, the expression on Buzz Aldrin's face has changed. Exhausted, but serene, after the most extraordinary journey ever undertaken by mankind, the astronaut rests in the LM shortly before it begins its ascent from the lunar surface.

'The surface is fine and powdery. I can kick it up loosely with my toe. It does adhere in fine layers, like powdered charcoal, to the sole and sides of my boots. I only go in a small fraction of an inch, maybe an eighth of an inch, but I can see the footprints of my boots and the treads in the fine, sandy particles.' The astronauts began their lunar EVA (Extra-vehicular activity) by setting up various experiments in the dusty plain, such as a seismograph and a system of reflectors designed to send back laser beams fired from Earth and thereby obtain accurate measurements of the Earth-Moon distance. Another experiment, this time a European one from the University of Bern in Switzerland, consisted of a hanging foil that would collect solar wind particles. They then set up a plaque to commemorate the event which Armstrong read out aloud: 'Here men from the planet Earth first set foot upon the Moon July 1969 A.D. We came in peace for all mankind.' Their last task was to collect twenty kilos of lunar rocks and dust and return to the *Eagle*. So far from home, it seemed to them a marvellous oasis amidst the boundless mineral expanse. To the south, they could see the resplendent Earth fixed in the sky, its blue tints contrasting with the grey shades of the little planet they had come to explore. A little more than two hours after Neil Armstrong's appearance on the steps of the lunar module that had marked the beginning of their Moon walk, the two men were back inside their spacecraft.

The time had come for the more serious business of returning home. Their golden beetle had a mass of 15 tonnes and measured 7 metres tall. It comprised two stages. The lower stage consisted of the descent engine with its now empty propellant tank, and the four metal landing gears. The upper stage was pressurised and contained the cockpit. It had a liquid propellant engine, with a thrust of only 1.5 tonnes, equivalent to that of a small aeroplane, that was nevertheless sufficient to take the module into orbit given the Moon's weak gravitational field. This was the ascent stage that would carry them up to redock with Michael Collins' command and service module, still orbiting at about 100 kilometres altitude. Before doing so, however, the two men were supposed to sleep. In the cramped confines of their craft, they lay down on the floor and shook with cold. After sleeping fitfully in the Earthshine for a few hours, they fixed on their harnesses and prepared to press on the button that

■ THE THICK LAYER OF DUST KNOWN AS THE LUNAR REGOLITH THAT COVERS THE MOON'S SURFACE HAS BEEN LAID DOWN GRADUALLY OVER MORE THAN THREE BILLION YEARS. THIS FINE LUNAR DUST IS PRODUCED BY THE CONTINUAL BOMBARDMENT OF COSMIC RAYS AND MICROMETEORITES.

would fire the ascent engine. At least, they hoped it would! And what if it did not? But in the spirit of the astronauts of the day, there was no place for doubts of this kind. They were all fighter or test pilots with thousands of hours of training behind them, programmed to react to the most critical situations without panic. In any case, they had not forgotten that, as a safety measure, the cabling that controlled the LM lift-off had been tripled. Less than twenty-four hours after the LM had touched down on the Moon, Armstrong's right forefinger pressed firmly on the ignition. The ascent stage took flight, abruptly, violently pushing away the thin insulation panels of the LM and leaving the great golden beetle poised for all eternity on the Moon's surface, with its four footpads placed firmly in the lunar dust.

RENDEZVOUS IN LUNAR ORBIT

In their module hurtling through the void at 7000 km/hr, Armstrong and Aldrin could see far away the tiny point of light that was the command and service module. It grew gradually brighter and soon, on a common orbit at 100 kilometres above the Moon's surface, the two craft got into line, approached one another and gently docked together. When Michael Collins opened the hatch that separated him from his two fellow travellers, it was two excited children he welcomed on board. In a spirit of jubilant buffoonery, Neil Armstrong and Buzz Aldrin exuberantly congratulated one another, overjoyed at having brought the mission off successfully. They were now well on their way home. Seen from there, the Earth seemed so friendly and hospitable, but so far away.

On 24 July, the three men splashed down in the pacific. They were impatiently greeted by over a billion people who had been following their exploit on television and radio. But in Houston and Cape Canaveral, the engineers and technicians were already preparing a new Saturn V rocket and training its crew. After the near religious fervour engendered by the Apollo 11 mission, it was in a mood of almost total indifference that Charles Conrad, Richard Gordon and Alan Bean left for the Moon, four months after their illustrious predecessors. And yet Conrad, piloting the lunar module *Intrepid*, brought off a quite astonishing feat when he made a pinpoint landing with his craft only 200 metres from the American unmanned probe Surveyor 3 that had softlanded there two and a half years previously. This time the landing site was just south of the immense Ocean of Storms (Oceanus Procellarum), in a vast plain located a hundred kilometres or so from the great Lansberg crater. When Charles Conrad and Alan Bean first surveyed the surrounding area from the hatch of the LM, they could clearly make out the probe, completely intact, with its three metal footpads sunk into the lunar regolith. Its long sample-taking scoop was stretched out, permanently motionless, above a few scattered rocks. What an extraordinary rendezvous! The two astronauts gazed up in awe at the two solar arrays reaching out a good metre above their heads and giving Surveyor 3 the appearance of an extraterrestrial robot. They walked several times around this genuine representative of lunar archaeology before taking a sample of its outer casing that would later be analysed back on Earth. The aim of NASA planetary scientists was to measure the effects of micrometeorite bombardment to which the lunar surface is continually subjected. For, unlike the Earth, the Moon has no thick atmosphere to keep such objects away. Conrad and Bean also recovered its camera, a somewhat difficult task. Having done this, they used plastic containers to collect over thirty-four kilos of lunar rock and dust. After four hours exploration, it was a simple matter to carry all this back to *Intrepid*. In the Moon's weak gravitational field, this material weighed only six kilos. Houston engineers and technicians felt that they could now take this kind of mission in their stride: the faultless lunar lift-off, rendezvous in lunar orbit, the breathtaking cruise across the void that separates Earth and Moon, then re-entry via the Earth atmosphere aboard the little Apollo capsule and finally a gentle splashdown in the Pacific Ocean with the help of three enormous parachutes.

APOLLO 13 ON THE BRINK OF THE ABYSS

The Apollo 13 mission took off six months later, on 11 April 1970. It was to land at the Fra Mauro crater at the edge of the Ocean of Storms. On 13 April James Lovell, Jack Swigert and Fred Haise were already well on the way to the Moon, at more than 300 000 kilometres from their mother planet. Now blasé, Houston technicians were following the

■ DURING THE APOLLO 17 MISSION, GEOLOGIST HARRISON SCHMITT LOOKED FOR CLUES AS TO THE MOON'S MYSTERIOUS ORIGINS. IT SEEMS THAT THE MOON IS THE DAUGHTER OF THE EARTH AND A PLANET THAT HAS SINCE DISAPPEARED. THE LATTER, THE SIZE OF MARS, HIT THE EARTH SOME FOUR AND A HALF BILLION YEARS AGO.

spacecraft's long and quiet passage as it began gently to curve towards the Moon. Suddenly Jack Swigert's voice rang out crisply over the loudspeakers with the most striking euphemism in the whole history of space travel: 'Houston, we have a problem.'

Indeed, the Apollo 13 craft, so far from Earth, was on the point of falling into the infinite abyss of space. A few moments before Swigert's call, the crew had been surprised by a major explosion. The little space fleet consisting of the LM and the command and service module had begun to tremble, then pitch from one side to the other. The situation was critical, alarms were sounding, red lights flashed almost everywhere on the control panels. Distraught, James Lovell looked through a porthole and observed: 'It looks to me, looking out the hatch, that we are venting something. We are … we are venting something out into the … into space.' The three men were confronted with a terrible situation and faced up to it with quite exemplary self-control. Helped by ground control, the crew soon understood what had happened. The violent noise they had heard was due to the explosion of an oxygen tank in the service module, ignited by a spark from a short-circuit. The incident was as simple and catastrophic as the one that had lost the crew of Apollo 1 three years earlier. Around the vessel, the consequences of the explosion were spectacular. A cloud of liquid oxygen that had immediately frozen as it left the shattered tank had spread out across space, and resembled a kind of wild snowstorm. Soon the perfectly spherical cloud was more than thirty kilometres across. As the spacecraft imperturbably continued along its path towards the Moon, and was thus still moving away from the Earth, the crew quickly realised how serious their situation was. The liquid oxygen lost into space was destined to provide them with good air for breathing during the trip, but also, combined with liquid hydrogen, to supply electricity for the spacecraft via a fuel cell. However, the situation was even worse. The walls of a second oxygen tank in the service module had been ruptured by the explosion of the first and its contents were beginning to leak out. Busy communicating with Earth, where ground control was doing everything possible to reassure the crew, and checking onboard systems as they failed one after the other, the three men fortunately did not have the time to think about what was

happening to them. They nevertheless understood that they were in mortal danger. Deprived of a large part of its energy reserves, *Odyssey's* service module had begun to give up the ghost. But it was this very module with its powerful engine that was supposed to bring the Apollo 13 crew back to Earth at the end of the mission.

The crew and Houston engineers were to put together an unprecedented space rescue. The service module had to be considered as lost since its electricity supply would soon run out completely. The idea was therefore to use the LM *Aquarius* as a kind of lifeboat. No sooner said than done. James Lovell, Jack Swigert and Fred Haise shut down all the command module's onboard systems and switched on those in the LM. The crew would now obtain their electricity and oxygen supply from the lunar module. Corrections to the trajectory of the drifting spacecraft would also be carried out using the LM engine both before and after lunar flyby. Despite their misfortune, some good luck was to come their way: to cater for the possibility of incidents after translunar injection, NASA engineers had calculated a trajectory for the Apollo missions that would allow the craft to return naturally to Earth after half a lunar orbit. In fact, Apollo 13 had already made its normal midcourse correction, which would take it out of a free return to Earth trajectory and put it on a lunar landing course. A first lunar module burn successfully pushed the

craft back onto a free return course. If everything went according to plan, the Apollo 13 astronauts would automatically return to Earth in precisely three days. But would they still be alive? The situation aboard the spacecraft in distress was not improving. The air renewal system in the command module had broken down and the astronauts were continually exhaling carbon dioxide into their micro-atmosphere, in quantities that the LM could not cope with. Another flashing light began to indicate that they would soon die of asphyxiation. Without panicking, carefully following the invaluable instructions of Houston technicians, the astronauts put together a system for coupling an air filter to the LM ventilation system. Saved for the time being, other problems were to occur on their fated journey. After the lunar swing-by, and with the tiny and distant Earth once more in view, flight control announced that their trajectory would no longer bring them back to Earth. The continued venting of oxygen from the damaged tank had altered their orbit and they might well be lost forever in space. They would have to make a small correction using the LM engine, even though it was not at all designed for such a manoeuvre. However, the lunar module proved to be remarkably adaptable. In a few seconds, Jim Lovell had reset the spacecraft onto the right trajectory. Now the captain and his two crewmen would just have to sit tight for a little over two days of discomfort and solitude. Aboard the lunar module

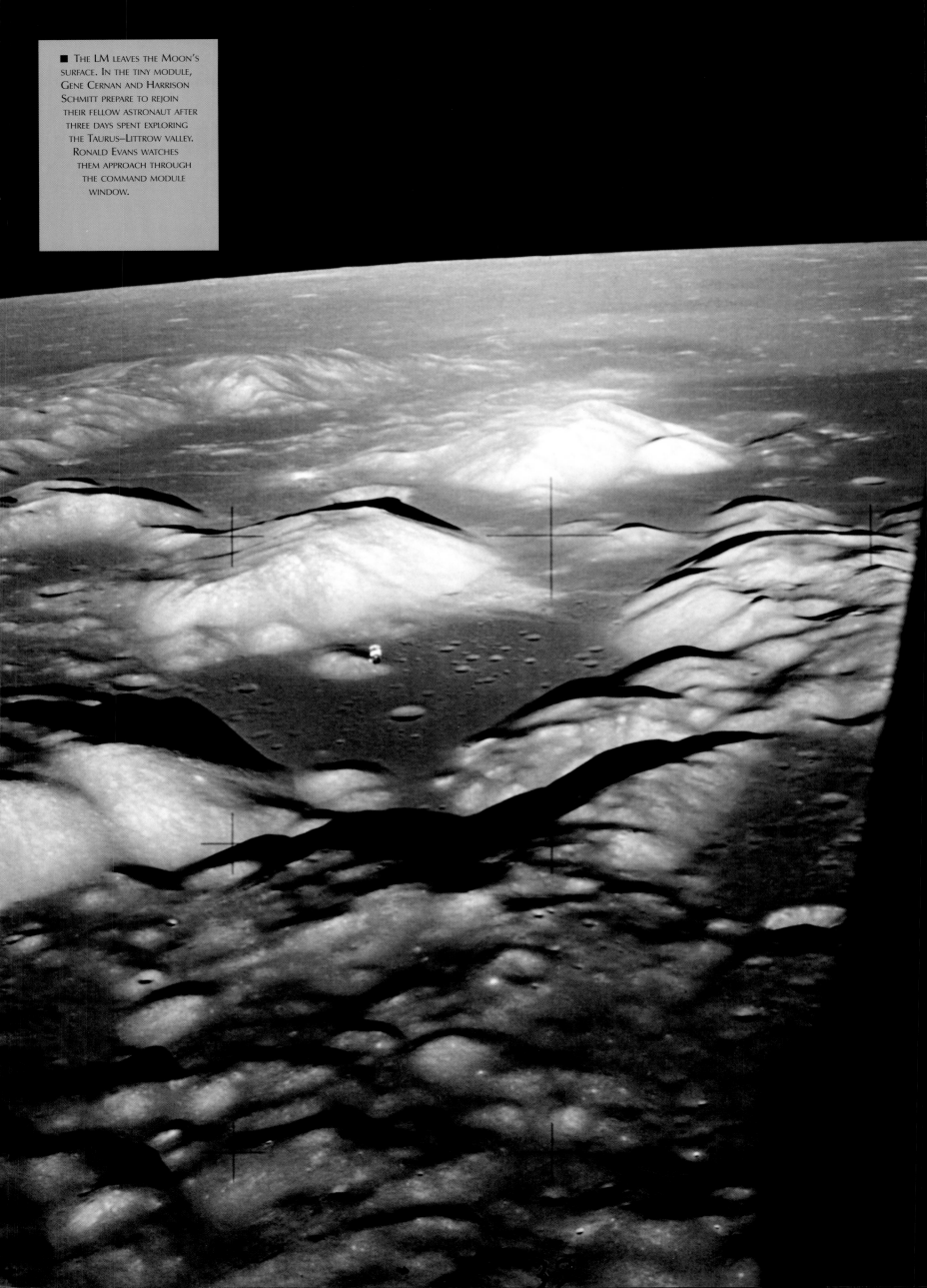

■ THE LM LEAVES THE MOON'S
SURFACE. IN THE TINY MODULE,
GENE CERNAN AND HARRISON
SCHMITT PREPARE TO REJOIN
THEIR FELLOW ASTRONAUT AFTER
THREE DAYS SPENT EXPLORING
THE TAURUS–LITTROW VALLEY.
RONALD EVANS WATCHES
THEM APPROACH THROUGH
THE COMMAND MODULE
WINDOW.

Aquarius, survival conditions were growing almost unbearable. Without proper recycling of the air, water vapour breathed out by the three men was condensing everywhere. The walls of the craft were gleaming with moisture. The temperature on board had plummeted and the three men, fastened into their soaking space suits, were shivering at only 3°C. To make things worse, they were very concerned by the state of the computer systems. Although the astronauts were freezing, the computers deprived of adequate ventilation, were beginning to heat up to a dangerous level, and threatened to break down.

Nevertheless, despite the terrible incident, Apollo 13 turned out to be an astonishingly robust machine. When the Earth had grown once more to fill their portholes, Jim Lovell, Jack Swigert and Fred Haise could prepare for re-entry through the atmosphere. In order to do so without mishap, they would have to bring the *Odyssey's* service module briefly back to life in order to resupply the command module with energy. Hurtling through space at nearly 40 000 km/hr, the crew were relieved to find that all systems switched back on perfectly after their long hibernation. They only needed the service module for a few minutes. Then they said a long goodbye to it through the porthole as its metal carcass drifted away. Finally, when the command module was ready, the three men squeezed for the last time through the airlock that led from the LM and closed the hatch behind them. *Aquarius* had saved their lives. It was with a certain emotion that they triggered the jettison. Stretched out in the cabin as it fell vertiginously towards Earth, they could feel once more the incredible force of gravity taking hold of their bodies. They would soon return safely home.

For NASA, the epic of Apollo 13 had been a successful failure. Although a lunar landing had not been possible, this was considered a minor detail. The American space agency had managed to bring its astronauts safely back to Earth in conditions never before encountered. However, disaster had only just been avoided, a reminder to all concerned that space missions are complex and risky. The engineers realised that they could not expect to send men to the Moon on their Apollo spacecraft without one day meeting with some terrible accident. At the outset, the

■ ON THE MOON, THERE IS NEITHER WIND NOR RAIN. THE ONLY EROSION IS CAUSED BY THE INCESSANT BOMBARDMENT OF MICROMETEORITES AND HENCE THESE FOOTPRINTS WILL REMAIN UPON THE LUNAR SURFACE FOR MILLIONS OF YEARS.

Apollo programme had scheduled ten missions to the Moon. Given that the competition with the Soviet Union had ended in a technical knockout and that public interest had faded even before the Apollo 12 mission, the purely scientific benefits were clearly insufficient to justify the exorbitant costs involved. NASA wisely decided that they would simply cancel the last three missions originally scheduled, Apollo 18, Apollo 19 and Apollo 20.

FOOTPRINTS IN THE LUNAR DUST

On 31 January 1971, less than a year after the Apollo 13 epic, another Saturn V rocket took off from Cape Canaveral. The Apollo 14 mission had been entrusted to Stuart Roosa, Alan Shepard and Edgar Mitchell. After an uneventful journey, they landed the lunar module *Antares* within 150 kilometres of the Apollo 12 touchdown. The Fra Mauro landing site, like those of Apollo 11 and 12, was only remarkable for its flatness. The LM came down in a vast plain of grey dust at the edge of the Ocean of Storms, about twenty kilometres north of the old Fra Mauro crater, a huge arena measuring ninety-five kilometres across that had been partially buried under an ancient lava flow. Around the LM, Alan Shepard and Edgar Mitchell forty-two kilos of rock samples and set up many scientific experiments. They spent a day and a half on the Moon before returning to Earth.

It was only with the Apollo 15 mission that the Apollo programme took on truly epic dimensions. Up to this point all lunar missions had aimed at vast plains with as little relief as possible for reasons of safety. In many ways these were the least interesting targets from a geological standpoint. But when David Scott and James Irwin opened the hatch of the lunar module *Falcon*, the landscape they discovered was quite breathtaking. Having left Cape Canaveral on 26 July 1971 with fellow astronaut Alfred Worden, who was now orbiting the Moon in the command and service module *Endeavour*, their pinpoint landing amidst one of the Moon's most rugged terrains proved to be an extraordinary feat. Keeping a sharp eye on the altimeter, James Irwin had guided them diagonally across the Sea of Rains and slowly brought them towards the Moon's biggest mountain range, the Apennines. Tracing out a circular arc 600 kilometres long, this chaotic ridge was made by a collision with a gigantic meteorite some four billion years ago. The hills and mountains viewed by the two astronauts as they gazed at the horizon through the LM windows were nothing but a huge embankment thrown up by an impact worthy of Armageddon. Indeed, our Moon might have been blasted into a thousand pieces by such a blow. Irwin had landed *Falcon* at the foot of Mount Hadley and Mount Hadley Delta, two peaks with deceptively gentle outlines, reaching up 5000 metres above the surrounding plain. Once outside, dazed by the timeless majesty of the landscape, Scott and Irwin's first task was to detach the first human-driven vehicle ever carried to another planet. This was the lunar rover, an electrically powered vehicle with a maximum speed of 20 km/hr. (Lunakhod 1 was the first ever Moon car, in November 1970, but it was unmanned.) They travelled about ten kilometres across the lunar surface, confronting hills and craters with astonishing ease. It was to be one of the most wonderful missions in the history of space travel. Like Tintin and his companions in Hergé's famous story, the two American astronauts discovered one surprise after another as they moved across scenery of quite unexpected beauty. Not far from the landing site, Scott and Irwin began by visiting the Hadley rille, a deep, winding crevasse that astronomers had been observing for decades with Earthbound telescopes. They stopped at the edge of the chasm, photographing the gently rounded and majestic slopes of the Apennines in the background. They were struck by the fact that it was impossible to determine the scale of these imperturbably serene hills. After a three-day sojourn on the Moon and having collected seventy-six kilos of lunar rock and dust, the two astronauts went back up to join Worden in orbit.

Veteran John Young, whose career in space already made impressive reading, left Earth on 16 April 1972 with Thomas Mattingly and Charles Duke in the direction of the Descartes landing site, one of the most rugged terrains on the Moon. In their lunar rover, they travelled for almost three days across the rough landscape, pitted with craters, picking up more than ninety-five kilos of lunar samples during the twenty-seven kilometre exploration.

RETURN TO THE MOON?

On 7 December 1972, when Apollo 17 took off with Harrison Schmitt, Gene Cernan and Ronald Evans aboard, the astronauts knew that

■ ON 14 DECEMBER 1972, THE MOON GROWS SMALLER AS THE APOLLO 17 MISSION DRAWS TO AN END. DID THE RETURNING ASTRONAUTS REALISE THAT THIRTY YEARS LATER, NO ONE WOULD HAVE RETRACED THEIR STEPS?

they were the last crew to be programmed for the Moon. They did not know whether someone would one day follow in their path. But would they ever have imagined that, thirty years afterwards, at the beginning of the twenty-first century, no one else would have set foot on the lunar surface? More than any other astronauts before them, Harrison Schmitt and Gene Cernan took full advantage of their stay on another planet. For the first time, one of the crew was neither test pilot nor fighter pilot. For the geologist Harrison Schmitt, every step on the Moon was one of wonder and discovery. Having left Ronald Evans to the solitude of his lunar orbit, Schmitt and Cernan landed the lunar module *Challenger* in a very rugged and ancient region of the northern hemisphere, the Taurus–Littrow valley, to the north of the Sea of Tranquillity. It was a wonderfully beautiful site. For three days, Schmitt and Cernan travelled over thirty-five kilometres through the foothills and around craters in the lunar rover. They discovered gigantic boulders displaced by ancient moonquakes and took over 115 kilos of rock samples from the ashen lunar dust. The landscape was reminiscent of the Atacama Desert in Chile. The apparently measureless hills that rose in stages above and beyond the vast plain seemed to defy the laws of perspective. Beneath their feet, the so-called regolith, the fine mineral dust resulting from billions of years of micrometeorite impacts, gradually changed colour with the relative position of the Sun. There were rocks everywhere and in the starless sky, black as coal above the ashen landscape, the Sun's glare was intolerable, only softened by the gentle presence of a strange blue planet, both incongruous and wonderful.

On 14 December 1972, Gene Cernan closed the hatch of the LM behind him. In under an hour he and his fellow moonwalker would rejoin Ronald Evans in lunar orbit. Their colleague had spent the past three days alone in the command and service module *America* and was probably smiling to himself at the thought of an impending opportunity to stretch his legs! Once the burn had been triggered and the ascent stage was on its way up into space, debris from the shattered thermal insulation layer and disturbed dust gradually settled around the blackened footpads of the LM. The lunar surface all around the abandoned vehicle would now remain frozen in its mineral immobility, carrying its testimony that bipeds had once been that way, visitors from another planet. These footprints would remain for a considerable time, until eventually wiped away by the relentless blast of cosmic rays and micrometeorites.

■ AT THE BEGINNING OF THE
THIRD MILLENNIUM, ASTRONAUTS
ARE LIMITED TO ENCIRCLING THE
EARTH AT LESS THAN 500
KILOMETRES FROM THE EARTH'S
SURFACE. ONLY THE TWENTY-FOUR
MEN WHO HAVE MADE THE TRIP
TO THE MOON HAVE BEEN LUCKY
ENOUGH TO WITNESS THIS VIEW
OF THE DISTANT EARTH,
SPLENDID BUT FRAGILE, A
TINY BLUE PLANET SWAMPED
BY THE INFINITE
BLACKNESS OF SPACE.

Ambassadors

■ A SPACE PROBE IS ABOUT TO LEAVE EARTH FOREVER. MAGELLAN HAS BEEN PLACED IN EARTH ORBIT BY THE SHUTTLE *ATLANTIS* AND WILL SOON DEPART FOR VENUS. AFTER A CROSSING OF SEVERAL MONTHS, THE PROBE WILL GO INTO ORBIT ABOUT EARTH'S SISTER PLANET. FOR THE MOMENT, VENUS IS NOTHING BUT A BRIGHT STAR A HUNDRED MILLION KILOMETRES AWAY, VISIBLE JUST ABOVE THE PROBE.

■ A LAST GLANCE BACK FOR THE GALILEO PROBE WHICH LEFT EARTH ON 18 OCTOBER 1989. GALILEO WENT INTO ORBIT AROUND JUPITER IN 1995 AND WAS STILL OPERATING IN 2001.

Seen from Earth, the Universe appears in all its infinite and impenetrable glory. The wonderful sky is scattered almost uniformly with stars and buckled around by the silver arc of the Milky Way, which forms a startlingly luminous belt. Far away, the tiny Sun is nothing but an unusually bright star, unable to sweep away the darkness of night. The boundless cosmic abyss falls away on all sides, cold, empty, and dark. To date, four probes have departed towards the four cardinal points and are now hurtling through the 'silence of this infinite void', which so terrified the French mathematician and philosopher Blaise Pascal. To all appearances, they seem to be perfectly immobile in a void that is no longer set to their scale. Indeed, as the decades go by, their escape from the Solar System is almost imperceptible. They have left the Sun, Mercury, Venus and the beautiful Earth–Moon couple far behind them. Four cosmic sailboats, launched across the waves of space, never to return. They cruised past Mars, loitering a while in the vicinity of Jupiter and Saturn, before casting a glance towards Uranus and Neptune as they flew past. Then finally, they crossed the imaginary line that serves to bound the Solar System, the orbit of the last known planet, Pluto. And so they were gone.

They carry with them a small part of humanity. On board two of them, there is a gold plaque that sketches in very brief outline their own story and that of the life form that dispatched them as ambassadors out to the stars. The two others carry magnetic disks in their metal flanks with images and sounds of planet Earth engraved upon them: the testimony of human beings, musical masterpieces, the muted melody of the wind and the haunting song of the whale. These messages are not really intended for anyone in particular, but they had to be placed aboard because one day the two Pioneer probes and the two Voyager probes will be the most ancient surviving monuments ever erected by humankind.

In the summer of the year 2000, it had been twenty-three years since the two Voyager probes untied their Earthly moorings, and twenty-seven and twenty-eight years for the Pioneer probes. But their long haul across space has barely begun.

As their names imply, it was the two Pioneer probes that first opened up the trail to the stars. Pioneer 10 and 11 left Earth aboard Atlas–Centaur rockets on 2 March 1972 and 5 April 1973, respectively, headed towards the outer Solar System via the asteroid belt – it was not known at the time whether the latter represented a collision threat for space transportation – followed by the giant planets Jupiter and Saturn. In fact, Pioneer 10 carried out the very first exploration of Jupiter, and Pioneer 11 the first exploration of Saturn. Although they did not revolutionise our

On 5 May 1989, the American probe Magellan is gently removed from the payload bay of the shuttle *Atlantis*. Here, its solar panels are still folded down. A few moments later, its engines will thrust it towards Venus, where arrival was scheduled for 10 August 1990.

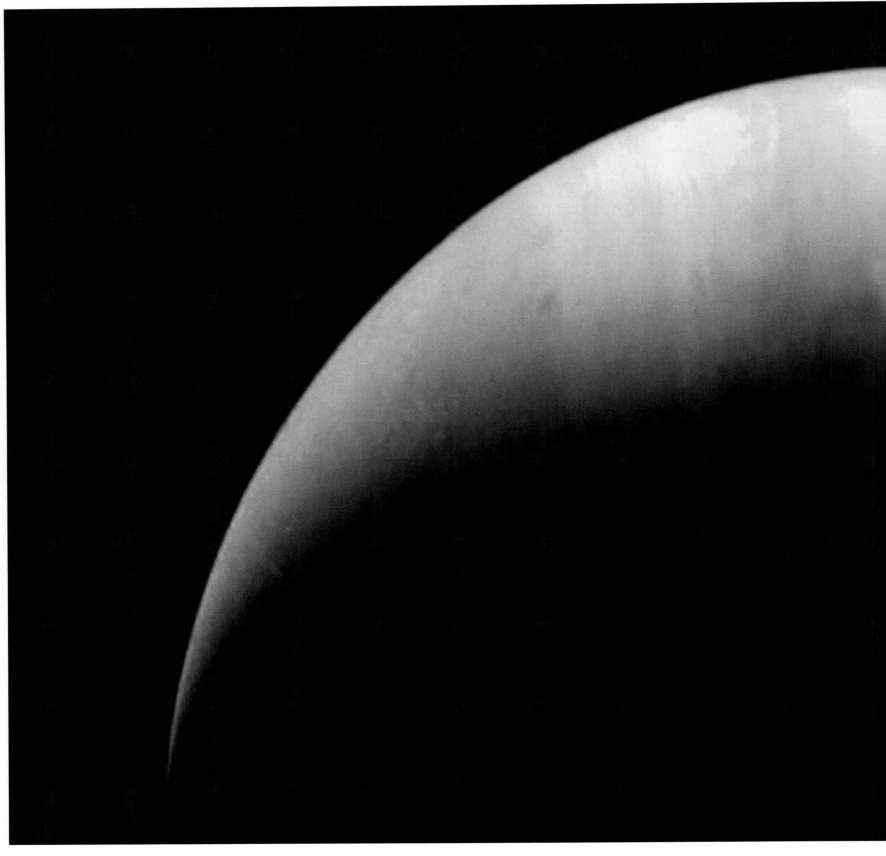

understanding of the Jupiter and Saturn systems, the Pioneer probes did allow American scientists and engineers to prepare for future, considerably more ambitious deep space missions. It was barely four years later, on 20 August 1977, that Voyager 2 began its great journey aboard a Titan–Centaur rocket, launched at Cape Canaveral in Florida. Like its sister Voyager 1, which it preceded by just a few days, it is a genuine technological wonder compared with the small and unsophisticated Pioneer probes, or even compared with today's space probes. Pioneer represented the last variation on the theme of 1960s technology, whilst Voyager prefigured the innovations of the 1980s. The fates of the two space caravels, so different and unexpected, their unbelievable exploration, perhaps the most fruitful in the whole history of

science along with Charles Darwin's *Beagle* expedition, are now living symbols of twenty-first-century astronautics. It is safe to say that they will one day take their place in human mythology.

Discovering the Solar System

The achievements of the Voyager 2 probe and the engineers who guided it from one world to another in our Solar System are quite extraordinary to recount. Like Voyager 1, the probe first crossed Jupiter's miniature planetary system, revealing as it passed the turbulent cloud formations of the giant planet and indeed filming them with an unprecedented abundance of detail. Its cameras

SYSTEM IN MARCH 1979, TAKING ONE OF
THE MOST BEAUTIFUL CHIAROSCURO
PORTRAITS OF THE SOLAR SYSTEM'S BIGGEST
PLANET.

then turned towards each of the planet's four Galilean moons. These owe their name to the great Italian physicist who first observed them in January 1610 with a refracting telescope of his own making. Until the Voyager flybys, these four satellites, of comparable size to our own Moon or slightly larger, appeared as tiny points of light even using the most powerful telescopes available to astronomy. The majority of specialists, being less imaginative than nature, expected to find satellites resembling the Moon. But these were worlds in their own right that came to light under the scrutiny of the two passing probes, planets of ice or fire, with unique landscapes of unsuspected beauty. To begin with, there is Io, a little globe that is continually being worked

over by the tidal tug-of-war between its gigantic neighbour and the three other Galilean satellites. Its surface changes all the time under the influence of tremendous volcanic eruptions which fling plumes of molten sulphur as high as 300 kilometres, whilst elsewhere black lakes of lava lap gently on their shores. Further out from Jupiter in its orbit, Europa is a much calmer world, looking rather like a ball of ivory. Its surface is a single, all-encompassing ice floe, the largest in the Solar System. The almost pure ice is sometimes veined with thin streaks of light brown rocks and the surface is criss-crossed all over with thousands upon thousands of cracklike features. Here and there, the ice floe looks just like the Arctic Ocean in spring, when the ice begins to break up and drift away. Europan-style icebergs several

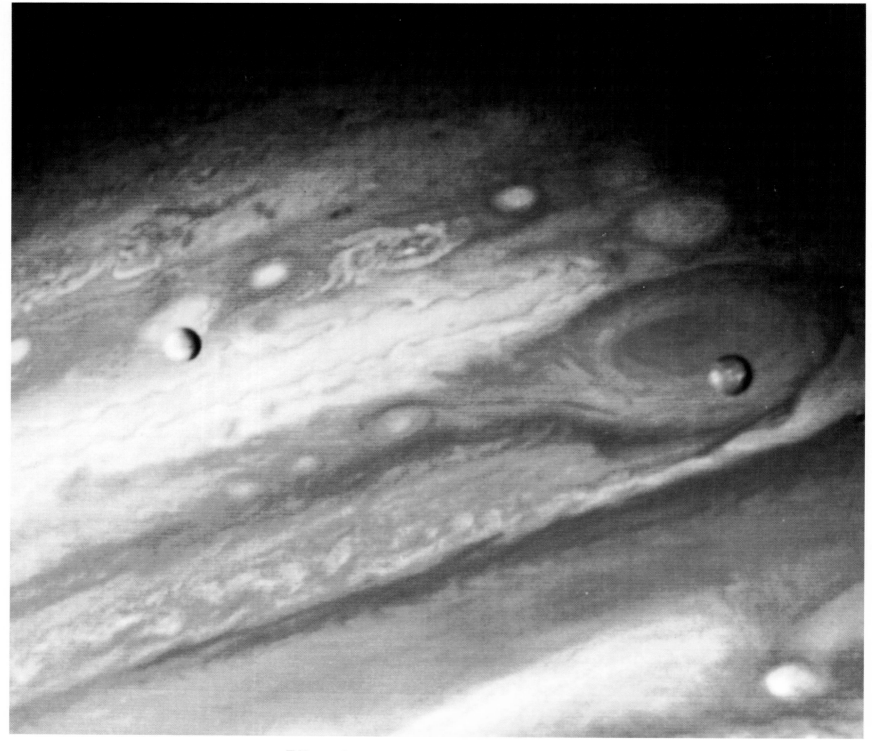

kilometres across jostle one another in a quite indescribable chaos. Just after the Voyager flybys, planetary scientists suggested that Europa's icy surface might conceal a vast ocean as much as a hundred kilometres deep. Confirmation of this wonderful hypothesis had to wait another twenty years, when the Galileo probe arrived in the Jovian system. Unlike its illustrious predecessors, whose work it was to pursue, this probe would remain for five years in the neighbourhood of the giant planet. Still more excitement was to come when Ganymede was revealed to insatiable planetary scientists. This is the largest natural satellite in the Solar System, much larger than the Moon, and harbouring a wealth of interesting geological features. For example, there are strange grooved terrains, where millions of square kilometres are covered by patches of more or less parallel ridges and wrinkles, juxtaposed in an apparently haphazard way. In the end, only Callisto, the fourth of the Galilean satellites, came anywhere near to fulfilling astronomers' expectations. Revolving far from Jupiter and freed from the gravitational constraints that heat and distort its three largest neighbours, Callisto presents an almost uniformly cratered surface, a Moonlike landscape that has probably looked more or less the same for the past four billion years.

AN UNEXPECTED WEALTH OF WORLDS

Leaving the Jovian system at more than 50 000 km/hr, the Voyager 2 probe curved towards Saturn, arriving in August 1981, a few months after the more rapid Voyager 1 which had overtaken it. And like its predecessor, it discovered a proliferation of new worlds, each more weird and beautiful than the last. Indeed, Saturn controls a whole system of moons and rings, sometimes coexisting in a strange symbiosis. Once again, the planetary scientists went into raptures before this grand spectacle. Saturn's rings had been known since Galileo first observed them. At the beginning of the space age, astronomers had distinguished five

zones of different brightness and density in the splendid cosmic hula hoop that swirls around the giant planet. But no one would have imagined that there would be tens of thousands of these rings, or rather ringlets, gravitating within Saturn's sphere of influence. As fine as worked gold, they are sometimes variable, sometimes highly eccentric, or distorted by the presence of one or two tiny moons within their midst. This astonishing roundelay was first observed from close up by the two Voyager probes, which also took the opportunity to examine the dozen or so moons already known to astronomers. Even there, the Voyager probes were to catch the specialists short, when they discovered that some of Saturn's satellites are worlds of pure ice.

Voyager 2 could have stopped there, proud to have accomplished its mission so perfectly, and taken a well-earned retirement. However, by an accident of celestial mechanics, combined with the reliability of the probe and the determination of American scientists, a quite different fate awaited the spacecraft. By a rare piece of good luck, the antepenultimate and penultimate planets of the Solar System, Uranus and Neptune, were destined to pass close by Voyager 2 as it continued its path across the heavens. NASA had known that such a propitious configuration would not occur for another two centuries and had thus programmed what they christened the Grand Tour. NASA engineers undertook a long and surrealistic dialogue with their spacecraft which was already more than a billion kilometres away. Their task was to change its onboard computer systems so that it could prepare for the new observing conditions it would have to face. Uranus and Neptune are much further from the Sun and hence much less well illuminated. This meant that they would be much fainter than Jupiter and Saturn. Everything on board Voyager 2 had to be adapted to the new conditions, including exposure times and tracking movements for camera shots and the rate at which data would be transmitted back to Earth, among other things. The encounter with Uranus took place in January 1986 at a distance

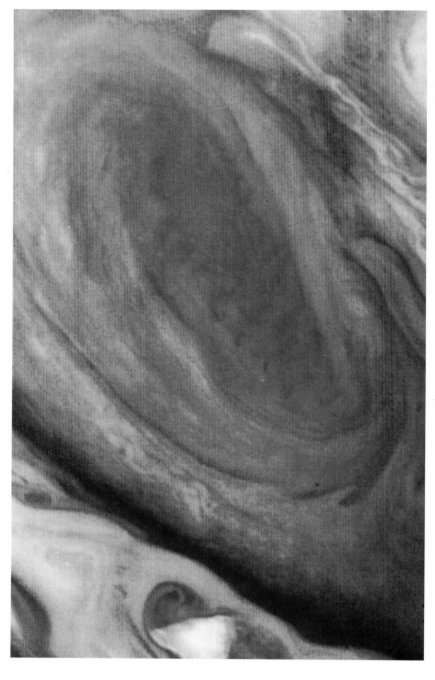

of 2.8 billion kilometres from the Sun and it was a memorable success. Apart from making wonderfully clear images of the ring system which surrounds the planet, Voyager 2 observed the moon Miranda and discovered an immense cliff, 15 000 kilometres high, separating two profoundly different geological regions. From Voyager data, planetary scientists were able to deduce that at some time in the very distant past, Miranda had suffered an impact of phenomenal violence with a gigantic asteroid. As it flew by the remote world of Uranus, Voyager 2 also brought to light ten new satellites.

VOYAGER 2 COMPLETES THE GRAND TOUR

Before the great leap into the infinite, there remained one last rendezvous, with Neptune, in August 1989. Prior to the Voyager 2 flyby, Neptune had appeared to astronomers as a tiny bluish marble. But Voyager was to discover a strangely beautiful world, a kind of cosmic 'big blue', a wonderful ocean blue atmospheric planet, overflown by brilliant, fast-moving, high altitude clouds and hollowed out by terrifying depressions where the most violent storms in the Solar System blow themselves out. Not far from there, Triton, the largest satellite in the Neptunian system, was found to possess a tenuous atmosphere. This thin shroud enveloped an astonishingly landscaped world of ice with liquid

nitrogen geysers dotted across the windswept plains. Under the gaze of Voyager 2, these geysers launched

■ UNLIKE MODERN PROBES, SUCH AS MAGELLAN, GALILEO OR CASSINI, WHICH WERE DESIGNED TO GO INTO ORBIT AROUND THEIR TARGET PLANET, THE VOYAGER PROBES MERELY

their ejecta into the scant airs where they immediately vaporised on encountering the freezing void which prevails at the confines of the Solar System. Such icy volcanoes had never been observed elsewhere. Before setting off on its journey of no return, which would take it forever from the Solar System that had spawned it, Voyager 2 sent back a final gift to its masters five billion kilometres away. Its cameras spied out five new moons and a delicate ring system within Neptune's gravitational sphere of influence.

Having passed through four planetary systems, Voyager 2 continued along its path, leaving so much data for the world's planetary scientists that, at the beginning of the twenty-first century, they have still not been completely analysed. It then sped away towards the stars.

Of the four ambassadors, Voyager 1 was the last to leave. On 5 September 1977, the probe was expelled from Earth's gravitational field by the tremendous force of a Titan–Centaur rocket. It very quickly reached the first and then the second cosmic velocity. Accelerating still further until it attained the extraordinary speed of 52 000 km/hr, it crossed the Moon's orbit after less than ten hours, then set off for a sixteen-month journey across the Solar System in the direction of Jupiter. It reached the

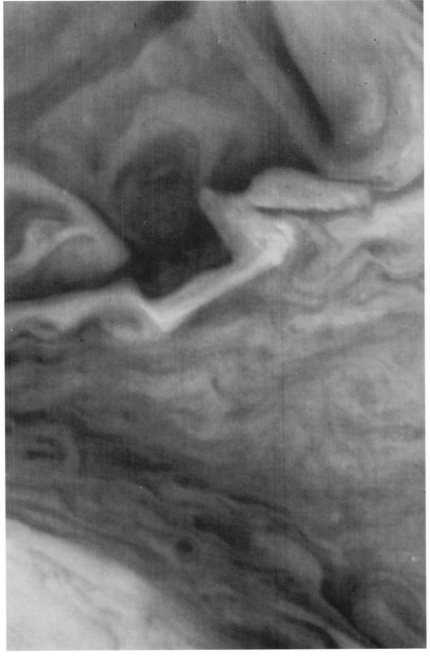

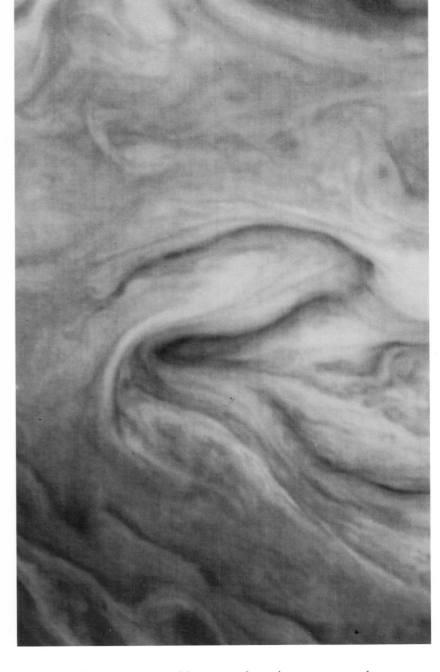

most massive planet in our system right at the beginning of March 1979, flying through the Jovian system in a few days, like Voyager 2 before it, and observing Jupiter, Io, Europa, Ganymede and Callisto from a different angle, with its onboard cameras, spectrographs and magnetographs. But Jupiter was only one stage in the mission of Voyager 1. Its main objective was a mysterious body, poorly known to the general public, Saturn's giant moon Titan. This distant object, bigger than our own Moon, has fascinated astronomers ever since they discovered that it had an atmosphere resembling that on Earth. On 11 November 1980, the probe entered the Saturnian system exactly on schedule. Its trajectory would cross Titan's at less than 4000 kilometres from the planet. Back on Earth, astronomers hoped to discover an even more exotic world than those revealed by the two Voyager probes when they flew past Jupiter. As it brushed past Titan's upper atmosphere, Voyager's cameras were expected to show the surface of the little planet with great accuracy, picking out mountains, canyons, craters, volcanoes and even oceans, should there be any. And once again, planetary scientists and probe pilots were in for a surprise … an unpleasant one! Indeed, what Voyager 1 discovered as it skimmed past the surface of Titan was that the planet was completely hidden beneath a perfectly opaque mask of fog. The surface was invisible.

However, the probe was able to confirm the presence of a terrestrial-type atmosphere made up of 90% nitrogen, methane and ammonia, but almost totally lacking in oxygen. The mission of Voyager 1 thus drew to a close. Beyond the mysterious Titan yawned the great gulf of interplanetary space, into which the probe was condemned to hurl itself at the record speed of 60 000 km/hr. Whilst Voyager 2 was directed towards Uranus and Neptune where it would continue its voyage of exploration and discovery, Voyager 1 would gradually catch up with and overtake the venerable Pioneer 10 and 11 probes to become, in February 1998, the most distant of humanity's stellar ambassadors.

During the Pioneer and Voyager sagas, the battle of the planets was raging back in the inner Solar System. On the blue planet, humankind had conquered the Moon and was dreaming of other worlds. Soviet and US engineers were vying to be the first to reach Mercury, Venus and Mars, the closest planets to Earth. It should be remembered that Jupiter and Saturn are basically fluid worlds, with no surface for a spacecraft to land upon. For example, gigantic Saturn is so light that, if there were an ocean large enough, it would float on the water. In contrast, Mercury, Venus and Mars are telluric planets, that is, they are much more like the Earth. Their mean density is that of the rocks that make them up and they are closer in size to Earth than to the giant planets. In addition, as far as Venus and Mars are concerned, they have their

■ IN NOVEMBER 1980, WHEN VOYAGER 1
CAME WITHIN RANGE OF SATURN, ASTRONOMERS
DISCOVERED A WORLD OF QUITE UNEXPECTED
COMPLEXITY. THE SUBTLE LIGHT VARIATIONS IN
THE WELL-KNOWN RINGS ARE DUE TO GRAVITY

own atmosphere, as has been known to astronomers for decades. This means that, like the Moon, these planets are reasonably accessible.

History will relate that Venus was principally a Soviet conquest, whilst Mars fell to the Americans. Before any space probe had ventured close enough to examine it carefully, Venus was generally considered as a kind of Earth-like paradise. The planet has the same size and mass as Earth and carries with it a thick atmosphere. According to some scientists, its inhabitants, if there were any, would enjoy a kind of never-ending tropical climate. For Venus is situated at only 110 million kilometres from the Sun, compared with 150 million kilometres for the Earth. In such idyllic conditions, it seemed natural to imagine exotic and maybe even evolved life forms on the Venusian surface. It was the American

probe Mariner 2 that first dispelled this illusion in 1962. Then, during the 1960s and 1970s, the Soviets made eight attempts to establish the truth about Venus. Venera 1 went silent shortly after launch, whilst Venera 2 ceased to operate before reaching the target planet and communications with Venera 3 were lost in its final approach. But with Venera 4 through 8, Soviet scientists felt they were getting warmer, and not just in the figurative sense! Physical conditions in the Venusian environment could hardly have been less hospitable. There was not the slightest trace of oxygen in the thick atmosphere, made up of 96 per cent carbon dioxide and less than 4 per cent nitrogen. By 1972, the year when Apollo astronauts were pacing out the lunar surface for the last time and NASA engineers

WAVES CAUSED BY TINY MOONS THAT ORBIT ON
EITHER SIDE OF THE RINGS. BECAUSE OF THEIR
CONFINING EFFECT, THESE ARE KNOWN AS
SHEPHERD SATELLITES.

were preparing to launch the two Pioneer probes, no one had any illusions about Venus: it was closer to hell than to heaven.

Venera 7 was the first probe to softland on another planet on 15 December 1970. It sent back very weak signals for twenty-three minutes after landing. Venera 8 repeated this performance, softlanding on 22 July 1972 and returning data for fifty minutes, a record in the dreadful conditions Venus imposes. This probe measured lighting levels which showed that photography would be possible. Venera 9 softlanded on 22 October 1975 and, although it survived only fifty-three minutes, was able to carry out successful TV photography. But how wrongly Venus had been considered as Earth's sister planet! When the probe finally touched down on the Venusian surface after a seemingly endless descent through the atmosphere, it lived a nightmare worthy of the best science fiction. The atmospheric pressure at the surface is ninety times greater than on Earth. In other words, going to Venus is rather like exploring the Earth's oceans at a depth of 1000 metres! The only difference is that on Venus there is not the slightest trace of either water or oxygen. The gentlest breeze would lift hypothetical visitors like straw in the wind and fling them afar as surely as any ground swell in the deep ocean. But the high pressure is only a beginning. The fatal blow for space probes visiting Venus is corrosion. What Venera 9 and its successors had to contend with was a continual downpour of sulphuric acid rain. And this leaves aside the temperatures prevailing under the planet's thick blanket of cloud. Before abandoning its data

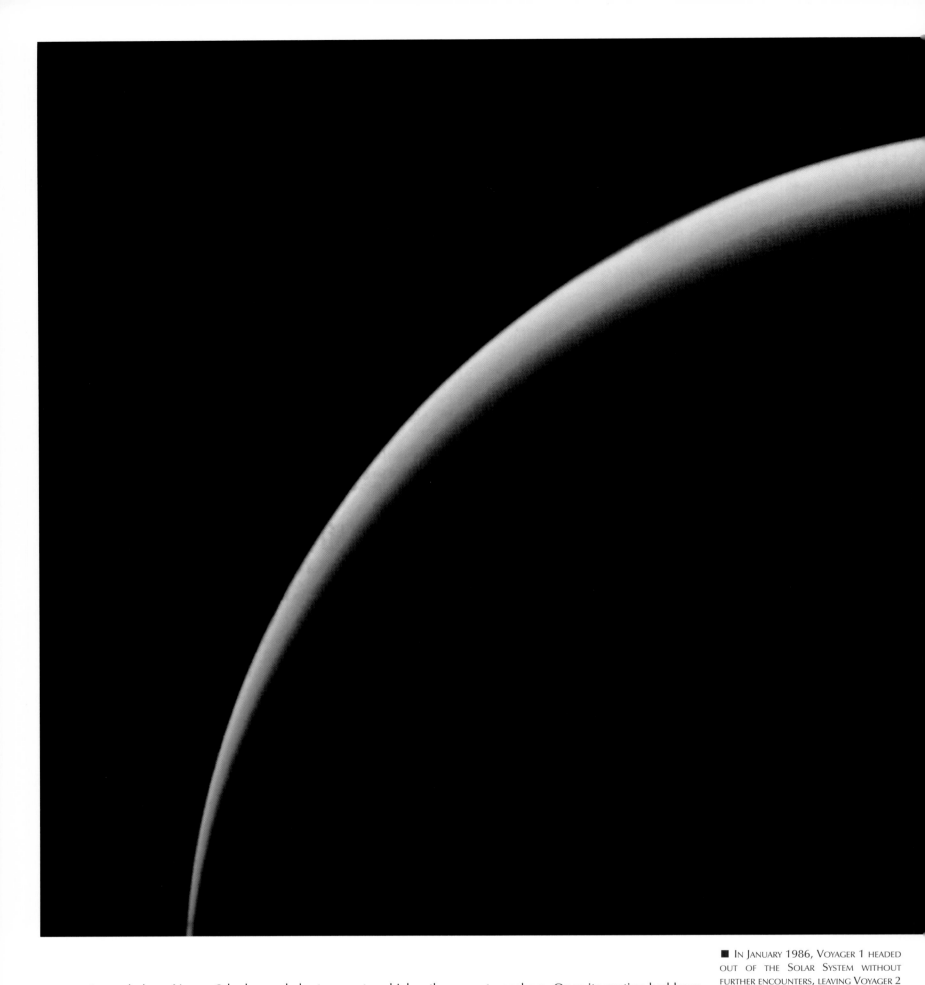

■ IN JANUARY 1986, VOYAGER 1 HEADED OUT OF THE SOLAR SYSTEM WITHOUT FURTHER ENCOUNTERS, LEAVING VOYAGER 2 TO CONTINUE THE GRAND TOUR ALONE. AFTER JUPITER AND SATURN, THE NASA

transmissions, Venera 9 had recorded a temperature higher than 460°C. What Earth-built craft could have survived such conditions? The Soviets were to persist, however, and they proved that they were up to the challenge. Venera 10, 11 and 12 prepared the ground for Venera 13 and 14, dispatched in the same launch window, and these sent back quite extraordinary panoramic colour views of the Venusian environment. When it penetrated the Venusian atmosphere at tremendous speed on 1 March 1982, the first of the two probes was protected by a thick heat shield rather like the manned Apollo capsule when it re-entered the Earth atmosphere. Once its motion had been sufficiently reduced by friction with the surrounding gases, a parachute opened. This happened at an altitude of around sixty kilometres where physical conditions are roughly comparable to those at sea level in the Earth's atmosphere. With the parachute open, the probe then fell slowly through the Venusian atmosphere, crossing ever denser and hotter layers of gas until, at an altitude of thirty kilometres or so, the parachute became unnecessary and was detached. Venera 13 continued its descent through the burning atmospheric sea and

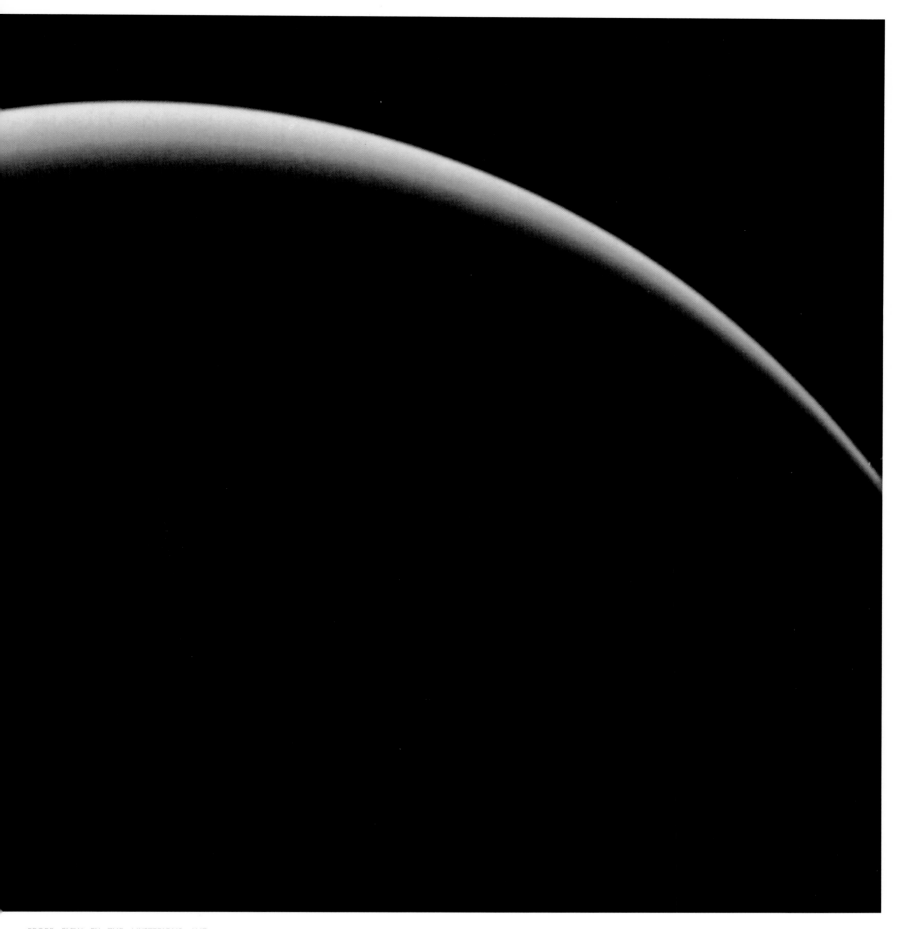

finally attained the planet's surface. The landing occurred not far from the equator, just east of the elevated region known as the Phoebe Regio site. Observing through a thick vitroceramic filter, the probe camera began to image a landscape that could have inspired Dante's Inferno. In the distance a diffuse horizon trembled beyond the furnace. On Venus, everything further than one kilometre away is blotted out by the thick atmosphere. In any case, the relief changes constantly as mirage follows upon mirage. In the uniform yellow sky, the Sun was invisible behind a cloud layer that never dissipates. There were no shadows to pick out the landscape, lit up by a brown glow comparable to a stormy day on Earth. The ground appeared to be made up of extremely fragmented and flat rocky plates, a mosaic of basalt slabs, scattered with small stones and some kind of gravel. The probe thermometer shot up to +470°C. Despite its protective hull worthy of a deep sea diving vessel, Venera 13 survived the pressure, acidity and temperature of the Venusian surface for only slightly longer than two hours. On 5 March 1982 Venera 14 landed less than 1000 kilometres to the south west, near the eastern flank of Phoebe Regio,

and resisted for slightly less than one hour.

After the achievement of the Venera probes, Soviet, American and French scientists buckled down to a more global study of the planet. The American Pioneer-Venus probes sent back superb images that showed the motion of the Venusian atmosphere, taken from orbit. Then on 11 and 15 June 1985, the two Soviet Vega spacecraft dropped two atmospheric probes into the atmosphere as they flew by on their way to observe comet Halley. These probes deployed parachutes like the Venera series, but for the first time in the exploration of the Solar System, they carried balloon probes amongst their payload. The balloons were designed and manufactured by scientists at the French Centre national d'études spatiales (CNES). They were launched at fifty kilometres altitude and swept away by violent winds gusting at over 250 km/hr, eventually covering a total of 11 000 kilometres in the Venusian atmosphere in only forty-six hours. Scientific experiments were loaded aboard little containers, the so-called Vega gondolas, weighing only seven kilos. These carried out accurate chemical analyses and studied meteorological conditions in this totally opaque upper layer of the thick atmosphere. The two Vega balloons recorded temperatures of 100°C and a similar atmospheric pressure to that at sea level on Earth. An extraordinary level of international cooperation was required to monitor in real time a balloon probe traversing the atmosphere of another planet a hundred million kilometres away. American and Soviet scientists worked together to collect data transmitted by the balloons, using their extensive space surveillance networks, including dozens of giant antennae scattered over the five continents, to relay findings to French research teams.

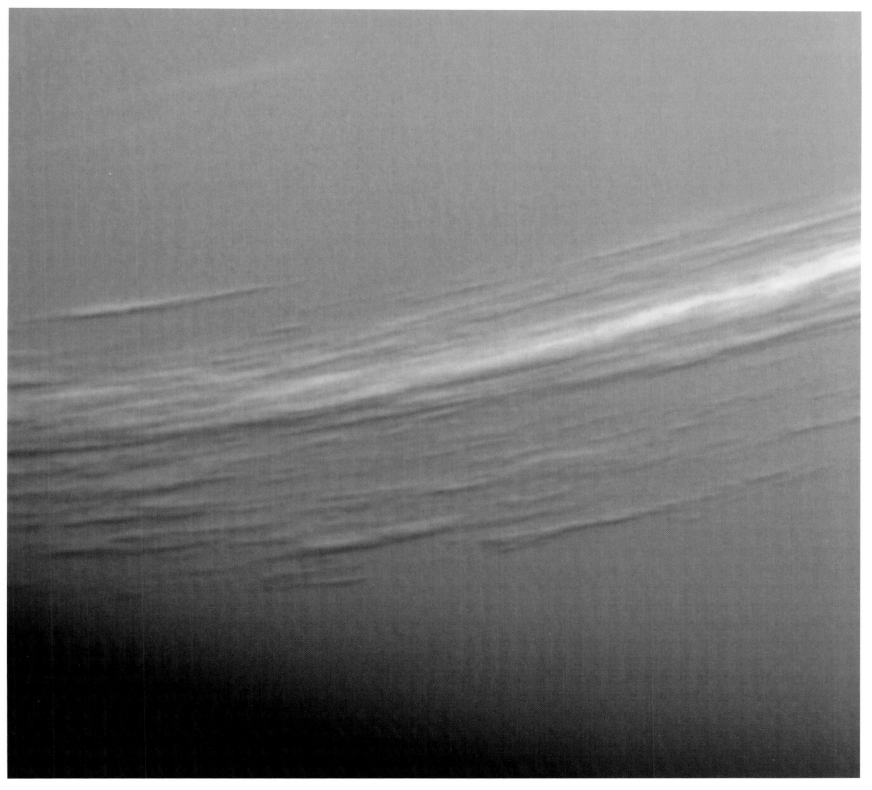

VENUS UNVEILED

■ PROBABLY ONE OF THE MOST BEAUTIFUL PHOTOS EVER TAKEN DURING THE EXPLORATION OF THE SOLAR SYSTEM. ON ITS POINT OF CLOSEST APPROACH TO NEPTUNE, VOYAGER 2 STUMBLED ACROSS A STRETCH OF HIGH ALTITUDE CLOUD, CASTING ITS SHADOW UPON THE LOWER ATMOSPHERIC LAYERS OF THE IMMENSE BLUE PLANET.

Once the ground and atmosphere of Venus had been revealed, it remained only to map the surface. By studying its relief, it would be possible to determine whether or not the planet was covered with impact craters like the Moon, or folded mountain chains and sporadic volcanoes like the Earth, witness to intense tectonic activity. In the case of Venus, it was impossible to use probes equipped with high-resolution cameras, like those still operating aboard the two Voyager probes far away in the outer Solar System. Such cameras could not see through the thick and uniform cloud cover that shielded the planet. Soviet and US engineers thus opted for radar. First, in October 1983, the Soviet probes Venera 15 and 16 were placed in orbit around Venus, followed by the US probe Magellan between 1990 and 1994. These produced very high quality radar surveys of the planetary surface. The Venera probes discovered a gigantic mountain chain, Maxwell Montes, reaching up almost 12 000 metres above the mean level of the surface – there is of course no 'sea level' on Venus! For its part, Magellan produced such superb radar images that they could have been mistaken for photographs, covering the whole surface of the planet. Let us just note in passing a constant feature in the history of space exploration: American space images, whether they be optical, radio, radar, infrared or X-ray, have always been sharper, and more beautiful to look at than Soviet or Russian ones. Could this be due to a cultural difference, or is it the result of an accumulated technological advance on the American side when it comes to state-of-the-art microcomputing and imaging? Whatever the reason, Magellan discovered a surprisingly yellow and relatively uncratered Venusian surface. Indeed, only a thousand or so impacts were recorded. The explanation is

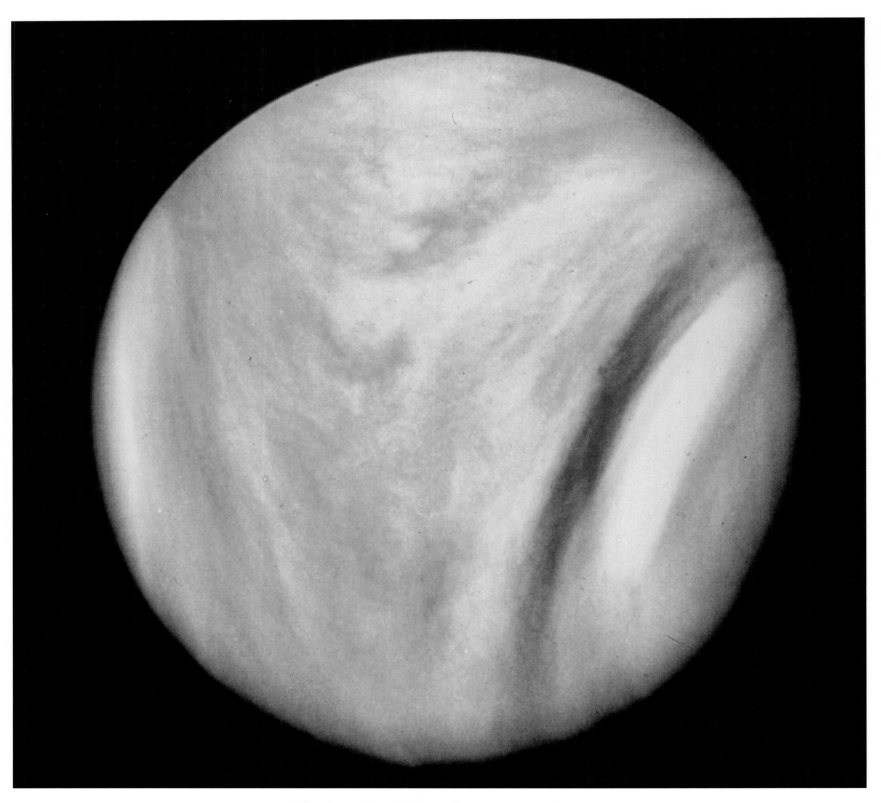

■ THE PLANET VENUS, PHOTOGRAPHED FROM ORBIT BY THE PIONEER-VENUS PROBE, IS COMPLETELY COVERED WITH CLOUDS AND MIST. ONLY RADAR DEVICES CARRIED BY THE VENERA AND MAGELLAN PROBES WERE ABLE TO SEE THROUGH THIS SEVENTY-KILOMETRE THICK CLOUD LAYER AND REVEAL THE PLANET'S VOLCANIC LANDSCAPES.

simple: for one thing, the surface of Venus is extremely well protected from small impacts by its thick atmosphere; for another, volcanic activity is the most efficient driving force for modelling planetary surfaces. In the desolate plains, most of which are a mere 500 million years old, Magellan revealed a thousand volcanoes in strange groups of volcanic domes associated with long, winding lava flows. The longest of these, Baltis Vallis, stretches out for over 6800 kilometres, the record for a planet in our Solar System.

MEANWHILE, BACK ON EARTH ...

On 18 November 1982, back on Earth, John Wilson yawned and glanced at the clock on the wall. In the half light, the digital display showed 0 h. It would soon be time to change target and trigger a new unmanned research programme. The American technician typed out a series of commands on his computer, causing a string of numbers to appear on the screen, whilst lifting a cup of hot coffee to his lips. Outside, the weather was fine as always. The scintillating stars shone harmoniously and with a rare intensity. For as the night had fallen at Goldstone almost six hours before, the wind had begun to strengthen. By now its cold and forceful blast was whistling through the steel girders of the 70-metre radio dish, obstinately pointed towards some invisible spot in the sky. At Goldstone the nights are harsh and a little disquieting for the occasional visitor, used to the bright tumult of town life. Located in California, in the heart of the Mohave Desert, Goldstone is a key element in NASA's Deep Space Network for monitoring transmissions from space. Indeed, its giant antenna, as noisy in the winter wind as a three-master buffeted in a storm off Cape Horn, follows all American, Russian or European space probes on their quests of

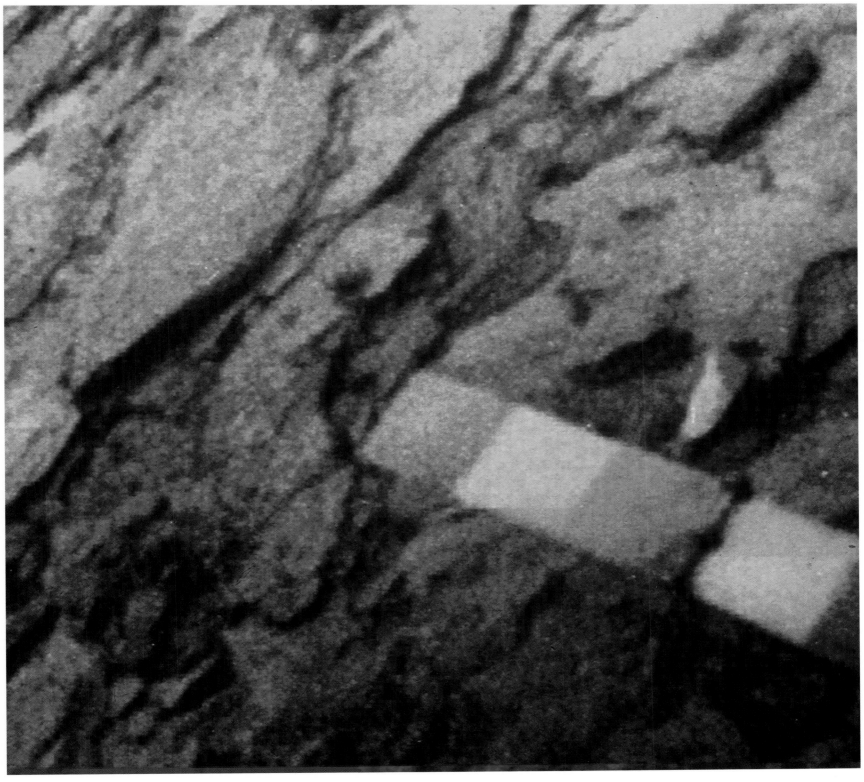

discovery. When the region of the sky in which a probe is situated passes below the horizon due to the Earth's rotation, the task is taken over by Goldstone's twin sisters at Tidbinbilla in Australia and Madrid in Spain. In this way, a given spacecraft can be accompanied day and night.

Obeying John Wilson's recent instructions, the huge dish began to swing round, rumbling like a large and gentle monster in a science-fiction film. Tilting meekly towards the east, it abandoned the tiny piece of dark sky where it had been patiently listening to data transmitted by the Voyager 1 probe as it crossed the distant reaches of space, so far from its homeland, engulfed somewhere between Saturn and Uranus. The seventy-metre dish soon picked out the brilliant Red Planet. Over there, two hundred million kilometres from Earth, the last vessel in a tiny scientific armada was still holding out, intently studying Martian weather systems and photographing the exotic but strangely familiar landscape of the Martian surface. For if Venus, as unveiled by Soviet exploration, had dampened enthusiasm in the domain of space exploration, given the hostile living conditions that would greet potential astronauts, it turned out that Mars, although not actually hospitable, was at least approachable and might even prove to be inhabitable. However, as in the case of Venus, the exploration of Mars had got off to a bad start for US and Soviet scientists. Naturally, each had sought to be the first to reach the Martian surface. Despite the relative ease of access, the world's two great space powers, and in particular the USSR, met with a quite improbable series of misadventures between 1962 and 1999. On the Soviet side, ten probes were lost: Mars 1, Mars 2, Zond 2, Mars 4, Mars 5, Mars 6, Mars 7, Phobos 1, Phobos 2 and then Mars 96, whilst Mars 3 only functioned for twenty seconds. On the American side, Mariner 3 and Mariner 8 were launch

failures, and three other probes disappeared: Mars Observer, Mars Climate Orbiter and Mars Polar Lander. As a result, the greater part of our knowledge of this planet comes from the American probes Mariner 4, 6, 7 and 9, Viking 1, Viking 2, Mars Global Surveyor and Mars Pathfinder. In 1964, Mariner 4 was the first probe to overfly Mars, sending back the first close-up pictures of another planet. Mariner 6 and 7 sent hundreds more pictures back in 1969, whilst in 1971, Mariner 9 was the first Mars orbiter and the first ever artificial satellite of another planet.

Making a distinction between launch failure and probe failure, the almost systematic Soviet problem with probe failures should be contrasted with the unprecedented reliability of the initial US probes. Of a total of seventeen successfully launched American probes, all took their missions to completion. The first US probe failure was Mars Orbiter in 1993, only occurring after an astonishing thirty-one trouble-free years of exploration. The USSR only ever managed to survey Venus, while the USA explored the whole Solar System with the exception of Pluto, an undoubted twentieth-century success story.

On 18 November 1982, the Viking 1 probe had been transmitting from the Martian surface for almost seven years. Its extraordinarily long lifetime could be explained not only by its

robust and reliable design, but also by the relatively clement Martian climate. Viking 1 functioned correctly for a total of 2280 days, roughly 50 000 times longer than the Soviet Venera probes in the inferno of the Venusian lower atmosphere. The two Viking probes left Earth on 20 August and 9 September 1975, going into orbit about Mars at the beginning of summer 1976. From there, each orbiter launched two landing modules, which softlanded on 20 July and 3 September 1976, respectively. Their main goal was to search for life. Viking 1 touched down in the Chryse Planitia region at latitude 22°48' north, whilst Viking 2 landed much further north, in Utopia Planitia at latitude 48°.

Landscapes discovered by the Viking television cameras were reminiscent of the Saharan reg or the Atacama Desert. The Chryse plain, scattered with rocks of volcanic appearance, stretched away to a featureless horizon. Here and there, boulders were partly buried in sand banks that sketched out small dune formations. Everything was covered with a fine dust and the whole scene was a brownish orange hue under the salmon pink sky. Towards the south-west horizon, an impact crater in the form of a large basin with a jagged rim could be made out through the mist. Above the crest line of a series of dunes on the horizon, the pink sky shifted to dark blue at the zenith. Further north, the Utopia Planitia site discovered by

■ A COMPLETE MAP OF THE SURFACE OF MARS WAS DRAWN UP BY THE TWO ORBITERS, WHICH ALSO CARRIED OUT METEOROLOGICAL STUDIES. IN THIS VIEW, THE SURFACE IS SCATTERED OVER WITH CLOUD BUT THREE VOLCANOES AND THE GIGANTIC CANYON OF VALLES MARINERIS ARE VISIBLE.

Viking 2 was even more monotonous, and harsher, too. The weather station deployed above the lander recorded more extreme meteorological conditions than in the Chryse plain. In the middle of the Martian winter, a fine layer of carbon dioxide snow falls in the desert, as temperatures plummet to –122°C. In such arctic conditions, Viking 2 nevertheless managed to hold out through two Martian winters before it finally ceased to transmit its daily report to Earth at the beginning of 1980. During almost seven years of operation endured by Viking 1, it recorded an absolute minimum temperature of –95°C whilst in the middle of summer the thermometer shot up to a heart-warming –20°C! The Martian sky, like that on Earth, is forever changing. The cloud layer is sometimes so thick that the Sun remains hidden and boulders on the surface no longer cast any shadow. However, the Martian atmosphere is as rarefied as the Venusian atmosphere is dense. On Mars, atmospheric pressure never exceeds 10 millibars, a hundred times less than at sea level on Earth. As a point of comparison, this corresponds to the atmospheric pressure on Earth at an altitude of 20 000 metres.

BLUES ON THE RED PLANET

In this light atmosphere, the probe recorded passing dust storms that darkened the sky, and photographed the effects of wind erosion at its landing site, in the form of small dust deposits here and there, slight changes in the surface albedo, and even a mini landslide at the base of a large boulder.

Whilst the Viking landers were registering these subtle changes in the Martian landscape, the two orbiters up in the sky of the Red Planet were systematically photographing the surface. Astronomers were fascinated by the world they revealed. The southern hemisphere is almost lunar, pockmarked by millions of impact craters. This landscape is three billion

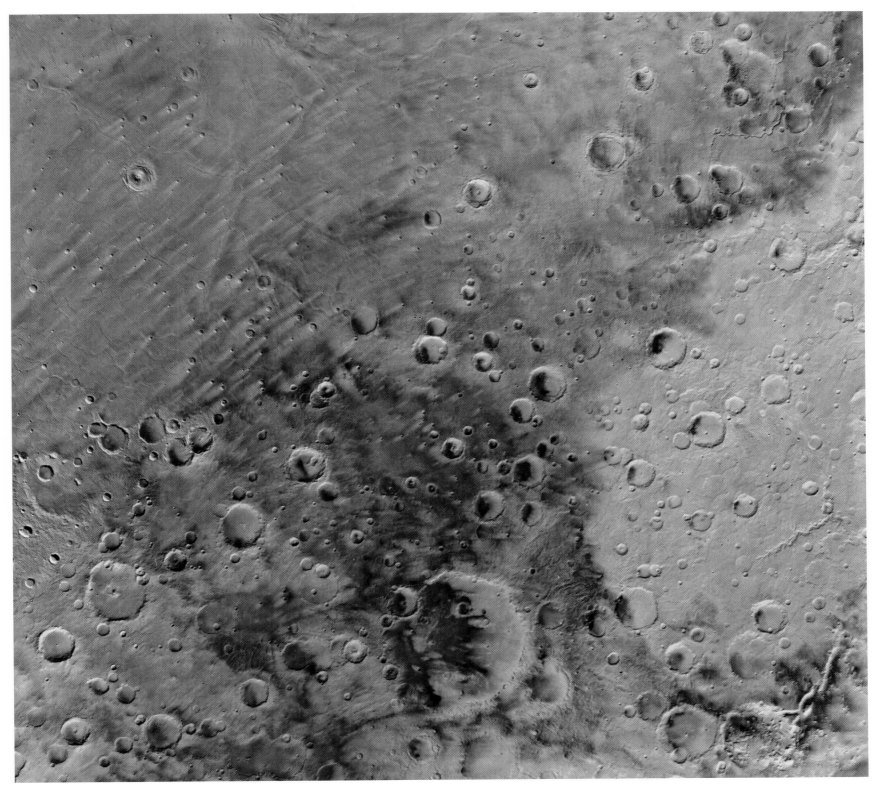

■ MARS CHANGES WITH THE SEASONS, JUST LIKE EARTH. HERE, IN A HEAVILY CRATERED REGION, MARTIAN DUST STORMS HAVE LONG SINCE BEGUN THE SLOW TASK OF ERODING IMPACT SITES MADE SOME THREE BILLION YEARS AGO. TRACES OF THE PREVAILING WIND ARE CLEARLY VISIBLE BEHIND THE RIM OF MANY SMALL CRATERS.

years old. The craters themselves are sometimes surprising. Inside one crater, a dune field may have formed in the shelter of the rim; inside another, the shock of the impacting asteroid may have literally liquefied the bedrock. Beyond, the landscape may have been partially concealed by passing lava flows. In contrast to this picture, the northern hemisphere is much less heavily marked by craters. Features in this region are dominated by the volcanic activity of a planet that is still living in the geological sense. To begin with, there are the four giant volcanoes of the Tharsis Rise: Olympus Mons, Pavonis Mons, Ascraeus Mons and Arsia Mons. Olympus Mons is the highest known peak in the Solar System, rising up 26 000 metres above the surface. Then there is Valles Marineris, a huge scar that stretches almost one third of the way around the planet's circumference, its impressive canyon walls rising up 6000 to 9000 metres above the surrounding terrain. At the two poles of the planet, as on Earth, the winter causes ice caps to form. Not far from the poles, there are immense dune fields, more extensive than the Sahara.

Each Martian year, in the frozen, windswept desert, dust storms are raised that sometimes cover the whole planet, hiding its surface from observation. But the most beautiful scenery discovered by the two Viking orbiters was perhaps the sinuous channels that look just like dried-up river beds. The specialists took their time to analyse the Viking photographs, but at the end of the twentieth century no doubt could possibly remain: water had indeed once flowed across the planet Mars.

In order to be quite sure of this conclusion, NASA planetary scientists sent another probe to the surface of the Red Planet, two decades after the Viking missions. Mars Pathfinder touched down at latitude 19°33' north at the mouth of Ares Vallis on 4 July 1997. Its camera, fixed at the top of a 1.5-metre metal mast, almost the height at which a man would hold it, produced a superb panorama of the Martian desert. Less than a kilometre

■ THE VIKING LANDERS GREW REMARKABLY WELL-ACCUSTOMED TO THE ARCTIC CONDITIONS PREVAILING IN THE MARTIAN DESERTS. VIKING 2 (ABOVE AND RIGHT)

from the landing site were two hills about forty metres high, called Twin Peaks. Beyond, the outer walls of an impact crater were clearly visible and further still, there stood a mountain about 450 metres high, some way towards the misty horizon. Ares Vallis is a flash-flood erosion feature, filled with a jumble of rocks deposited some three billion years ago by spectacular torrents of water or mud. Certain rocks seem to have been washed down and worn by the flow of water, whilst others are all leaning over the same way, as though pushed into this position by a strong current. Ares Valles is 1 500 kilometres long and about 100 kilometres across at the mouth, where Mars Pathfinder softlanded. It seems that volcanic activity may have melted underground ice reserves, thought to be plentiful in the Martian subsurface. The waterlogged ground then became unstable and collapsed, releasing gigantic but short-lived floods, a thousand times more abundant than those associated with the Amazon here on Earth.

All three of these Martian probes have now ceased to communicate, finally vanquished by the freezing climate of the Red Planet. The first was Viking 2 which stopped transmitting from the desolate Utopia Planitia in January 1980, after four years of faithful service. Then on 19 November 1982, the Chryse Planitia space station, alias Viking 1, also gave up the ghost. Mars Pathfinder resisted for almost three months before it too shut down on 27 September 1997.

Today, at the opening of the twenty-first century, John Wilson and his colleagues at Goldstone, Tidbinbilla and Madrid are still engaged in their patient monitoring of spacecraft that go in search of other worlds. After the first wave of discoveries made by the Pioneer and Voyager probes, the trend is towards *in situ* exploration carried out over longer periods. The aim is no longer a simple flyby. Instead the probe goes into orbit around its target object. Good examples are provided by Ulysses and SOHO, continually turned towards the Sun which they have been observing now for twenty-four hours a day since October 1990 and December 1995, respectively. Then there is NEAR, gravitating around the asteroid Eros, somewhere between the orbits of Earth and Mars, since the beginning of the year 2000. There is also the Mars Global Surveyor, patiently surveying the Red Planet since March 1999 and sending back images of quite exceptional quality; details on the photographs reveal features down to a few metres in size! But perhaps the most successful globe-trotter in this class is Galileo. It began with a flyby of the two asteroids Gaspra and Ida, during which it discovered Ida's moon Dactyl, the first known asteroid moon. It then went into orbit around Jupiter and dropped an atmospheric probe down through the clouds of the giant planet, before going on to produce a wealth of photographs of the four Galilean moons, Io, Europa, Ganymede and Callisto. It successfully withstood Jupiter's harsh radiation environment, going well past its planned lifetime. Having valiantly awaited the arrival of the Cassini probe, the two American probes worked in consort to send back stereographic observations of Jupiter before Cassini was carried on by its prearranged impetus, designed to take it to the planet Saturn for the year 2004. Once there, Cassini will release the European probe Huygens from its cargo bay. This module will finally realise a forty-year-old dream already dear to the designers of Voyager, namely, to explore Saturn's giant moon Titan and at last discover what lies beneath the thick cloud layer of this little hibernating version of Earth. Meanwhile, Galileo is still involved in Jovian satellite flybys, with Callisto in May 2001 and Io in August 2001.

CONTINUED OPERATIONS FOR OVER THREE
YEARS, WHILST VIKING 1 (LEFT) TRANSMITTED
BACK PHOTOGRAPHS AND WEATHER REPORTS
FOR CLOSE ON SEVEN YEARS.

TWELVE BILLION
KILOMETRES FROM EARTH

Secretly, all those guiding the great dish at Goldstone have a soft spot for the ambassadors. The youngest of them are the same age as these probes, and when these men and women go into retirement, they will hand over their care to others, passing on the precious equatorial coordinates needed to locate their thin and distant radio voices from far across the velvety sky. Of the four probes that have actually left the Solar System, one has already been lost. In November 1995, Pioneer 11 ran out of fuel and stopped transmitting at a distance of six billion kilometres from the Sun. Voyager 2 and Pioneer 10 are now located at nine and eleven billion kilometres from the Sun and are still monitored by the giant antennae of the Deep Space Network. Of the four ambassadors, Voyager 1 is now the furthest. It has long been travelling at the third cosmic speed, moving through space at a fantastic 62 000 km/hr. On a human scale, it is already unthinkably far away. On 1 January 2001, it had left the Sun some twelve billion kilometres behind it, or 31 578 times the distance from the Earth to the Moon. When it transmits back to Earth, the message takes eleven hours to make the trip at the speed of light!

On a cosmic scale, however, the probe has hardly even moved. We might even say that it is stationary. Four centuries ago, this is what Galileo had understood, that movement is nothing. Originally carried away at the speed conveyed to it by the powerful Titan–Centaur rocket twenty-three years ago, the probe needs no further energy input to keep it moving. It simply continues under its own impetus. It is likely that nothing will ever bring it to a halt. And yet it is to all intents and purposes immobile, since its speed relative to the heavenly bodies all around it is infinitesimal. Voyager 1's space odyssey has only just begun. As long as the onboard energy supply remains sufficient to run its telecommunications antenna, the Goldstone monitors will be able to keep track of it, a tiny signal amongst the cacophony of Earth-bound telecommunications satellites and the hullabaloo of new generation probes speeding through the Solar System in all directions. Needless to say, at twice the distance from the Sun to Pluto, the probe is no longer supplied by solar energy. In fact, mounted on the metal hull of Voyager 1 is a nuclear power generator whose core is composed of plutonium 238. This provides heat which is transformed into electricity to supply the various systems of the spacecraft. Among these, the radio transmitter is the probe's only remaining link with Earth and it has a power output of just twenty watts. Most of its emitted signal is lost in interstellar space, so that what reaches our Earth-based antennae is unbelievably weak. In fact, the power of the received signal is more than ten billion times weaker than the power output of the tiny battery in a digital watch. The engineers of the Deep Space Network know only too well that they will one day lose all trace of their probe. For one thing the received signal grows ever weaker as the probe moves further away, and for another the transmitted signal gradually declines as the plutonium 238 releases less and less energy. If bigger antennae are not constructed on Earth to track the tiny probe – and today astronomers have giant radiotelescopes measuring up to 300 metres across – the engineers at Goldstone will finally lose contact with Voyager 1 in 2020 when it reaches a distance of twenty-two billion kilometres, or 57 894 times the distance between the Earth and the Moon. However, there is every likelihood that they will feel inclined to keep in touch with it for another decade or so beyond this point. Only when its nuclear fuel has been exhausted will the probe finally fall silent, as it moves relentlessly away from our Solar System. In one century, it will be sixty billion kilometres from the Sun, and in 2000 years, it will

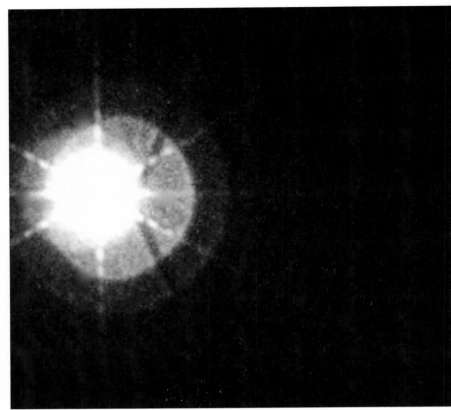
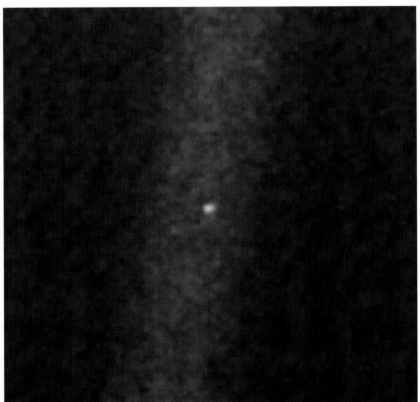

celebrate its first thousand billion kilometres. And yet this represents only the tiniest step. Despite the great distance, the Sun will still be by far the brightest star in the sky. In fact, Voyager 1 will take a great deal longer to escape completely from the Sun's sphere of gravitational influence. Not until then will it really begin to drift across the abyss that separates us from our nearest stellar neighbours. The solar escape velocity, also known as the third cosmic velocity, which only Voyager 1 and the three other ambassadors have yet attained, will only really prevail in 20 000 years from now, when Voyager 1 is one light year away.

VOYAGER 1'S LAST LOOK BACK

At this point, Voyager 1 will see only anonymous stars around it, shining in distorted and unrecognisable constellations. Even then,

its journey will have only just begun. This stellar sailboat and its three companions moving off in four different directions will embark upon a long interstellar crossing. Voyager 1 is moving towards the constellation of the Giraffe (Camelopardalis), Voyager 2 towards Sirius, south of Orion, and Pioneer 10 in the direction of Aldebaran in Taurus, which it will fly by at a considerable distance in about four million years. Finally, Pioneer 11 is heading straight for the centre of the Milky Way, in the region of the Eagle (Aquila) and Sagittarius. Their aluminium framework and their gilded magnetic disks will be cooled to around −270°C, the temperature of the interstellar void. Gently bathed in the few photons that reach them from the stars, occasionally grazed by a passing elementary particle or a microscopic dust grain that escaped long ago from the expanding envelope of some remote

■ ON 14 FEBRUARY 1990, BEFORE IT FINALLY LEFT US, VOYAGER 1 TURNED AROUND TO TAKE A LAST LOOK BACK AT THE SOLAR SYSTEM. AROUND THEIR STAR CAN BE SEEN THE DISTANT PLANET VENUS (TOP

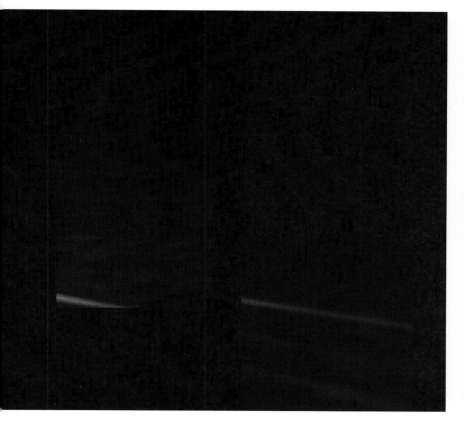

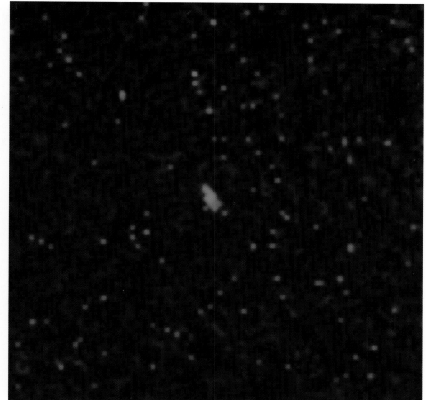

LEFT), THEN EARTH, JUPITER, SATURN, URANUS AND FINALLY NEPTUNE. TODAY THE PROBE IS LOCATED SOME TWELVE BILLION KILOMETRES FROM THE SUN.

supergiant, they will sit there as though in hibernation. No erosion will ever be able to affect them. Nothing will ever transform them.

The ambassadors' journey will never reach its end. They will witness the birth of new stars in the constellations of Taurus and Orion. In ten million years, they will see stars die that once animated the constellations of ancient Earthling legends. They will still be there when Rigel and Betelgeuse, Eta Carinae or Antares light up the whole galaxy for a brief instant before finally dying away. Every 250 million years, they will feel the tremendous ground swell of the galactic arm sweeping past them and fading again into the distance. They will wander forever on the great ocean of space, and longer again, revolving round the stellar

maelstrom that is our galaxy. In five billion years, they will witness the end of the Earth itself, brought about by the death of the Sun. From afar, they may drift silently past some other star with its invisible retinue of planets. In ten billion years, they will have travelled 500 000 light years, equivalent to a trip right around the galaxy, and yet a tiny step on the scale of the whole Universe. In this way they will move forever onwards, invisible and silent, until all memory of the ancient names of stars sparkling in our gentle night skies have been long forgotten.

They will remain out there forever, like four messages in four bottles thrown upon the mighty ocean of the galaxy, drifting across the unfathomable chasms that separate its thousand billion stars, ambassadors for all eternity of a forgotten civilisation.

■ THE ODYSSEY OF THE FOUR
AMBASSADORS HAS ONLY JUST BEGUN.
VOYAGER 1 IS HEADING TOWARDS THE
CONSTELLATION OF THE GIRAFFE
(CAMELOPARDALIS), VOYAGER 2 TOWARDS
SIRIUS, AND PIONEER 10 TOWARDS
ALDEBARAN IN TAURUS, WHICH IT SHOULD
REACH IN ABOUT FOUR MILLION YEARS.
PIONEER 11 IS MOVING TOWARDS THE
CONSTELLATION OF THE EAGLE (AQUILA)
IN THE HEART OF THE MILKY WAY. IN
TEN MILLION YEARS, IT WILL FLY BY
ONE OR OTHER OF THE STARS VISIBLE
IN THE BOTTOM RIGHT CORNER
OF THIS PHOTOGRAPH.

Escape from Earth

■ THE CARGO BAY OF THE SHUTTLE *ENDEAVOUR* OPENS NOISELESSLY. ABOARD THE CRAFT, AMERICAN ASTRONAUTS BOB CABANA, RICK STURCKOW, JERRY ROSS, NANCY CURRIE AND JIM NEWMAN AND THEIR RUSSIAN COLLEAGUE SERGEI KRIKALYOV WATCH THE *UNITY* MODULE SOON TO BE RELEASED FOR DOCKING WITH THE *ZARYA* MODULE. IT IS 4 DECEMBER 1998 AND THE GREAT ADVENTURE OF THE INTERNATIONAL SPACE STATION HAS JUST BEGUN.

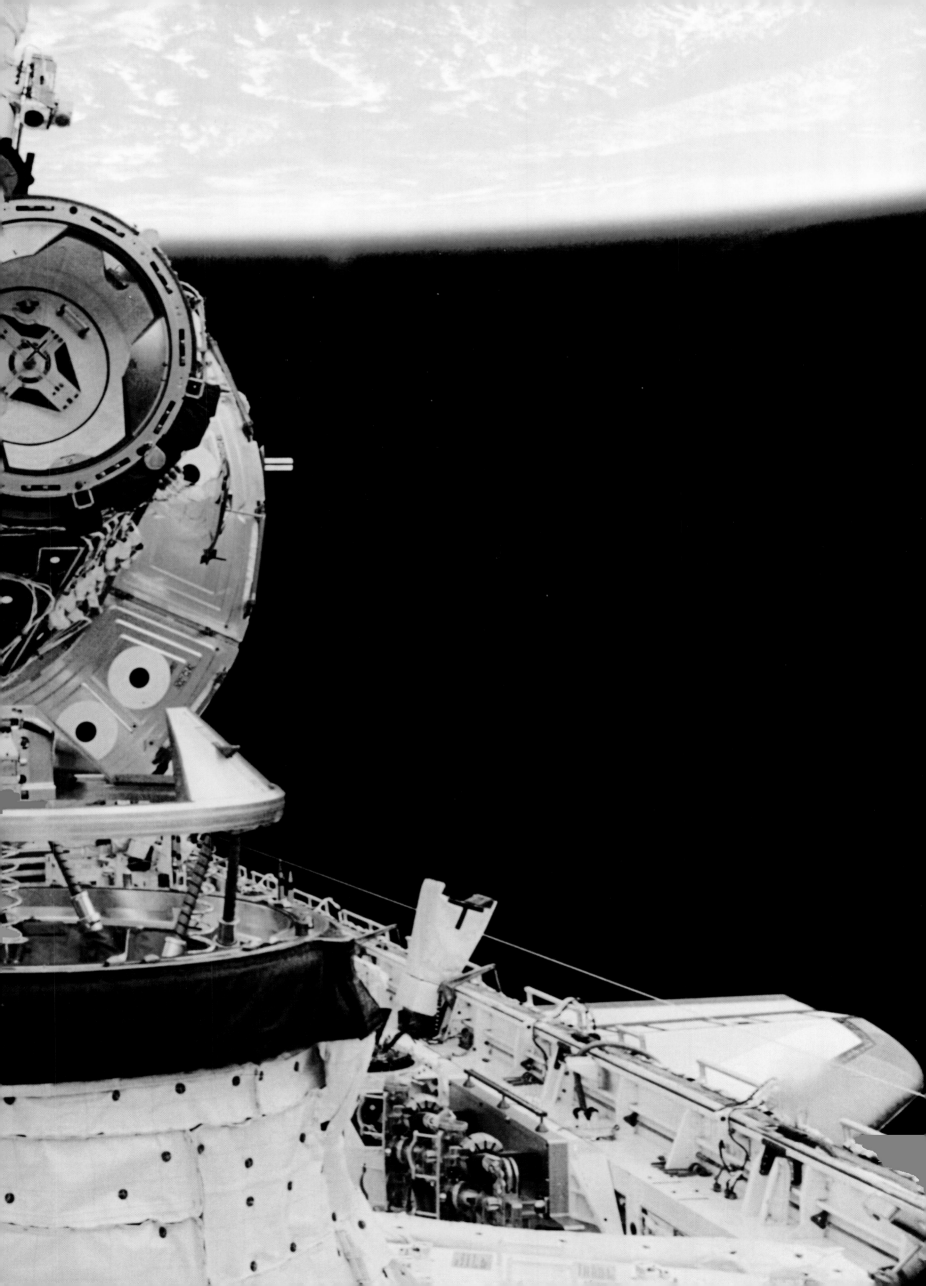

■ AMERICAN ASTRONAUT KATHRYN C. THORNTON ON A SPACE WALK.
AIR PRESSURE INSIDE THE SPACE SUIT MAKES EACH MOVEMENT OF THE
HANDS RATHER DIFFICULT.

The International Space Station (ISS) has been revolving around the Earth since 20 November 1998 on a slightly higher orbit than the one once occupied by Mir. Small and modest by comparison with its majestic predecessor, it is still only in its embryonic stages, growing in space at its leisure. Indeed, the new space station should have a long life before it. It is the first attempt at truly worldwide cooperation in the history of space conquest, bringing together nations and space agencies which had previously been competitors, or even outright enemies. The circular orbit of ISS at 400 kilometres altitude passes over both Baikonur and Cape Canaveral, the two main launch complexes destined to provide supplies over the coming decades. The first module of the space station was *Zarya*, launched from Baikonur by a Proton rocket. This pressurised cylinder looks very like the Mir modules. It measures 13 metres long and 4.3 metres in diameter and weighs about twenty tonnes. Deployed on either side are two gigantic solar arrays like two great silver wings. Less than two weeks after it was placed in orbit, it received a visit from the US shuttle *Endeavour*. On board were American astronauts Bob Cabana, Rick Sturckow, Jerry Ross, Nancy Currie and Jim Newman, and their Russian guest Sergei Krikalyov. In the cargo bay was the second module, *Unity*, which Jim Newman and Jerry Ross would dock onto *Zarya*. *Unity* is in fact a connecting module (or node) measuring 7 metres in length and 4.5 metres in diameter. It has six

docking ports to which future ISS modules will eventually be attached. Two further groups of astronauts visited the embryonic space station aboard *Discovery* and *Atlantis* on 27 May 1999 and 18 May 2000, respectively, in order to install various pieces of equipment and prepare it for its first genuine crew.

The first crew destined to actually live aboard the ISS, the so-called expedition one crew, was Sergei Krikalyov, Yuri Gidzenko and Bill Shepherd. Before they could fly from Baikonur in autumn 2000, it remained to attach the *Zvezda* service module, the early station living quarters, equipped with a life-support system and command and control facilities. This was the first fully Russian contribution and just like certain Mir modules, it comprised a docking port for Soyuz spacecraft. It was launched unpiloted on 12 July 2000 and successfully docked by remote control on 25 July of the same year, using the Russian automated rendezvous and docking system Kurs. Expedition one crew was launched on 31 October 2000 and docked 2 November 2000. They remained aboard for over four months, during which time ISS Assembly Flight 5A brought up the Destiny Laboratory Module, the centrepiece of the ISS, measuring 8.5 metres in length and 4.3 metres in diameter. This pressurised module is designed to house near-zero-gravity science experiments. The first crew returned on 21 March 2001 aboard the same flight that brought out the expedition two crew, James Voss, Susan Helms and Yuri Usachev.

■ The new NASA space suits are miniature worlds in their own right, with their own heated and pressurised atmosphere. Astronauts like Winston E. Scott, shown here during a space walk, are protected from the Sun's ultraviolet light by a thick visor which filters out ultraviolet radiation.

Proposed and directed by NASA, the ISS project has already brought together Russia, Europe, Canada, Japan and Brazil. It is a truly monumental undertaking, unprecedented in the history of space conquest, with the notable exception of the Apollo programme. Including the host of shuttles and Semiorka launch vehicles required to build up this gigantic space Meccano, module by module, the cost of the ISS can be estimated at around 100 million euros. This is on a par with the cost of Apollo. In terms of technical and human capacities, there are several considerations. On the American side, three out of the fleet of four shuttles will be mobilised to serve the station, probably for the next fifteen years. (It is not planned to use *Columbia*, the oldest model of the shuttle, owing to its lower performance compared with the others.) If all goes according to plan, about fifty shuttle flights will be required through to 2006, at a rate of around one flight per month. These will carry out equipment for the station, as well as supplies and crews. On the Russian side, more than twenty Proton launches have been planned and in Europe, the new heavy launch vehicle Ariane 5 will also be put to use. NASA's aim is to complete construction by the beginning of 2006. The result will be a genuine space village, ready to welcome astronaut teams. When finished, the ISS will comprise two large service modules with living quarters for the crew and six laboratory modules for scientific activities. There will be one EVA hatch to allow space walks and several docking ports to cater for the US shuttle, Russian Soyuz craft and European cargo and fuel supplies. Seen from afar, the ISS will look something like Mir, but on a quite different scale. Almost eighty metres long and weighing around 450 tonnes, it will be able to host crews of six or seven astronauts in a living space of some 1300 cubic metres. This is as roomy as a Boeing 747! Deployed on either side of the immense spacecraft, along a mast 108 metres long, will be eight pairs of solar panels supplying about 100 kW of solar power to the ISS.

So much for the construction itself. But what is the purpose of such a huge space complex? The main motivation for the ISS is above all geopolitical. For the Americans, the aim is to take the great technological nations, especially Russia, along the road to a large scale international collaboration. The Russian space industry was in danger of collapsing when the country found itself launched along the road to a liberal economy. In this context, NASA's order for Russian modules for the ISS came as a boost to the Russians but it was a gift that did not come without

■ THE ISS IS THE MOST AMBITIOUS TECHNOLOGICAL PROGRAMME OF THE THIRD MILLENNIUM. STILL IN ITS EARLY STAGES, IT WILL GRADUALLY GROW TO FULL SIZE IN 2006.

ulterior motives. By taking control of the ISS programme, the United States guaranteed themselves the lead role in space conquest for a long time to come. Furthermore, both Russia and Europe have been carried along and are now committed to making large contributions to what will essentially be an American space station. Indeed, some observers and politicians have already criticised the Russian and European share in the overall budget as unreasonable. For its part, NASA, whose budget exceeds that of all the other space agencies put together, will still possess the means to develop new launch vehicles and pursue its exploration of the Solar System.

THE GAMBLE OF THE INTERNATIONAL SPACE STATION

This said, a dozen or so countries have already signed up with NASA to take part in the ISS adventure. American, Russian, European, Canadian, Japanese and Brazilian astronauts will soon be carrying out fundamental research in its six laboratory modules. Just as happened aboard Skylab, Salyut, Mir and the laboratories carried by the shuttle, experiments on board the ISS will be largely devoted to gravitational physiology, space medicine and biotechnology. However, fundamental physical research will also feature with the study of crystal growth and fluid behaviour in microgravity conditions, and the development of new ultrapure materials and alloys. In addition small telescopes for ultraviolet and X-ray observations will also be set up on the central beam of the station. But in fact, as is often the case for fundamental research projects, many experiments to be set up aboard the ISS over the coming years have not yet been finalised and the corresponding instrumentation has not yet been designed.

The International Space Station is nevertheless a gamble that brings most specialists out in a cold sweat. The risks are enormous. One or more launch vehicles may fail in flight, losing a whole module or even an entire crew. NASA officials have not forgotten the 1986 *Challenger* disaster and take such risks very seriously indeed. Apart from the ninety space missions required to deploy and supply the ISS, NASA has also scheduled more than a thousand hours of space walks for its astronauts, in order to guarantee perfect control over the way the modules are pieced together. Even though these EVAs are prepared with the utmost caution, they remain highly dangerous missions.

From a detached standpoint, the ISS seems like a wonderful and fragile baroque edifice. The scores of return journeys made by the shuttle could have been avoided, and the multitude and complexity of the modules making up the ISS would not have been necessary if Russians and Americans had stuck to their development of conventional heavy launch vehicles. With the shuttle, an extremely costly and complex manned launch vehicle only capable of launching payloads of about thirty tonnes into Earth orbit, the Americans have little advantage over the unmanned Russian Proton rocket. The latter can put twenty tonnes into orbit at far less cost. And how can we speak of progress when we recall that twenty-five years ago when Skylab was the talk of the day, the old Saturn V was able to transport payloads of almost 100 tonnes into Earth orbit?

A PROCESSION OF SHUTTLE FLIGHTS

As far as the Soviet engineers were concerned, they had succeeded in developing their own extremely powerful launch vehicle, Energia, just before the break-up of the USSR. One of its main uses would have been to launch the Soviet shuttle *Buran*. Energia could muster 3 500 tonnes of thrust, even more than Saturn V, and could place masses of up to 100 tonnes into low Earth orbits just like its illustrious American counterpart. Unfortunately, it was only successfully tested twice, in 1987 and 1988, before being abandoned like Saturn V through lack of suitable missions. Considering today the difficulties involved in building the International Space Station, and the cavalcade of shuttle flights it requires, many engineers bitterly regret that the heavy-duty unmanned launch vehicles were dropped in favour of a costly manned space shuttle with much lower capacity. In just five Energia flights, the Russians could have placed in orbit a simpler, bigger and cheaper ISS. But would the Americans have asked them to participate in a space station in which they could no longer be the masters?

Scheduled for completion in 2006, after an eight-year construction period, assuming there are no incidents along the way, the International Space Station will long feature in the technoscientific landscape. Crews should follow on continuously one after the other, each flying for at least six months at a time. Apart from their scientific and technological activities, these crews will be living confirmation that the world's great powers intend to perpetuate human presence in space.

■ First check-up visit for the *Zarya* module. Crews of six or seven astronauts will be relayed out to the ISS where they will enjoy a living space of 1 300 cubic metres.

Visible across the whole planet, voyaging noiselessly through the dawn the world over, the International Space Station will be a small bright star gliding across the constellations, a dream in constant mutation as the speeches assign it the role of geopolitical symbol, scientific opportunity or token of technological excellence. But above all, like the ascension of some inaccessible Himalayan peak, it is perhaps a shared dream, an inexpressible but profoundly human desire, at the limits of sporting achievement and metaphysical quest.

As the ISS modules dock one by one, a second wave of Solar System exploration will also begin. We may recall that, during the last two decades of the twentieth century, most of the larger bodies of the Solar System were visited by unmanned space probes. By the year 2000, American, Russian and in some cases European probes could list visits to the Sun, Mercury, Venus, the Moon, Mars, Jupiter, Saturn, Uranus and Neptune, not to mention their natural satellites numbering about fifty in all. The European probe Ulysses deserves particular mention, the sole object yet placed in polar orbit around the Sun. Even comet Halley and four asteroids, Gaspra, Ida, Mathilde and Eros, had been engaged in close encounters. The only major objects to escape from this extraordinary inventory were the remote pair made up of Pluto and its moon Charon, which mark the known limit of our planetary system at almost five billion kilometres from the Sun. No probe has yet been near them. Once this preliminary campaign of observations had been achieved, planetary scientists were keen to deepen their knowledge of the various objects, revealing as they had a quite unimaginable diversity and complexity.

A genuine armada consisting of several dozen American, European and Japanese space probes will thus be sent forth in the direction of the Sun, Mercury, Venus, asteroids, comets, and Pluto. New exploration campaigns will also be launched towards the more interesting planets, starting with Mars, then Jupiter and Saturn. However, the idea is no longer to flash past the target like the earlier Pioneer and Voyager probes, but rather to set up house in the neighbourhood, or even on the surface of the planet in question.

Mars is the most obvious target. Close to Earth and easily accessible, requiring only about ten months to get there, it also happens to be relatively hospitable. It is not too massive, neither too hot, nor too cold, and holds about it a tenuous atmosphere vaguely resembling that of our own planet. Indeed it is a firm favourite with scientists, engineers

the ESA are beginning to believe what once seemed unimaginable or even impossible. The world's two great space agencies are working on what may become the finest scientific mission of all time: the search for extraterrestrial life. Following the unexpected discoveries at the end of the twentieth century, scientists have made a statistical estimate of ten billion for the number of planets revolving about stars in our own galaxy. Obviously there is a great temptation to investigate whether conditions on some of these planets might not be propitious for the development of life. From a technological standpoint, we are close to present limits, but astronomers have taken up the challenge. The idea is to set up a network of telescopes in space at something like the distance of Jupiter. Operating in concert, these would be sufficiently powerful to analyse the chemical composition of the atmospheres on all exoplanets resembling Venus, the Earth or Mars and located at the same distance from their star as these from our own Sun. Biologists and astronomers assert that, if life has appeared on one or other of these planets, it will be detectable through its interaction with the planet's atmosphere. For example, it should be possible to measure an ozone or oxygen excess, witness to biological activity, as it is on

Earth. These missions – Darwin for Europe and Terrestrial Planet Finder for the USA – are clearly without precedent in the history of humanity. The motivation to realise this millennium quest is probably sufficiently strong amongst scientists around the world to ensure that the two projects merge together in the years to come. Indeed, they are exceptionally complex and costly and the technological challenge is far-reaching. Darwin or its American counterpart could not expect to see the light of day before 2020 or 2025. Needless to say, if the thousands of planets conscientiously catalogued by astronomers should all prove to resemble Jupiter, Saturn or Venus, human enthusiasm for exoplanets would soon dwindle. Humankind would once again be faced with an overwhelming sense of solitude in the cosmos. But what if the great space telescopes of the future should spy out blue planets, with pure skies rich in oxygen and long coastlines awash with the waves of great oceans?

There would then be an irresistible temptation to go and take a closer look. But is this actually possible? At the present time there can be no doubt that the answer to this question is no. Over the past fifty years, we have grown accustomed to reading science-fiction accounts of interstellar travel. However, current or future rockets capable of navigating in Earth orbit or even across the entire

Solar System would be powerless if it came to crossing the vast chasm of space that separates us from the nearest stars. It seems that the Universe was not made for humanity's wanderings. Even with the best spacecraft available to date, it would take ten years or so to cross the Solar System from one side to the other. At the same speed – of the order of 50 000 km/hr – it would take about a million years to cross the immense void that lies between our star and its nearest neighbours. Naturally, we are free to dream of future technological advances, but to what end? What could we do with a spacecraft travelling a thousand times faster, speeding across space at 50 million km/hr, if the travel time to the closest exoplanets was thereby reduced to a mere thousand years? But there is much worse than this. There is a very basic reason why interstellar travel would be unrealistic, arising from the very principles of modern physics as laid down in Einstein's theory of relativity.

The problem relates to so-called relativistic speeds, close to the fantastic 300 000 km/s at which light crosses the vacuum, and the only ones compatible with the idea of interstellar travel. According to Einstein's equations, which have never failed to make accurate predictions over the past ninety years, such speeds can only be approached by supplying truly vast amounts of energy, and the speed of light can never actually be attained by a massive object because an infinite amount of energy would be required. In other words, engines capable of powering a spacecraft out to the stars in times comparable with a human lifespan just do not exist, either in reality or in the physicist's wildest dreams.

There remains one more law of nature, imposed by the very structure of space–time itself, which looks likely to pose an even more serious problem. This law states that if a spacecraft eventually acquired a relativistic speed, a significant fraction of the speed of light, the rate of physical processes experienced by its

occupants would no longer match the rate of the same processes occurring on Earth. Put another way, if one day a few centuries from now a spacecraft were able to attain such speeds, propelled by an as yet undiscovered energy supply, so that it could reach a nearby exoplanet within a human lifetime, its proper time would be irrevocably decoupled from terrestrial time. When the spacecraft had been flying across interstellar space for several decades according to the time experienced by its occupants, several centuries would have gone by back on Earth! There is no remedy for this phenomenon. On such cosmic distance scales, relativity teaches us that travel through space is inevitably accompanied by travel through time in the sense described here. If human beings find a way of sending manned spacecraft out to planets gravitating about other stars, they will also be sending them into the future with no hope of return or even contact until dozens of generations have gone by on Earth. What could be the meaning of such missions? Will we seek one day to colonise the planets encircling other suns? Will the whole Galaxy one day become our world? Such questions remain purely academic today. The solutions dreamed up for cosmic travel, such as fuelling engines with antimatter, curving space–time, using black holes as spatio-temporal gateways, are as poetic and laughable as those imagined by Cyrano de Bergerac in the eighteenth century. In order to travel to the Moon, he took flight using flasks of dew heated by the Sun. It may be that we will never hear the screeching call of the night train as it leaves for the stars.

Unless of course the voyage is virtual. The Hubble Space Telescope, although held in the firm grasp of a terrestrial orbit, is able to supply images of the planets in our own Solar System which are almost as detailed as those transmitted by space probes. Here, perhaps, lies the solution to realising humanity's starlit dreams.

Astronomers and space agencies may soon be inspired to think along these lines. Just as Mars Pathfinder allowed us to visit the landscapes of planet Mars from the comfort of our armchairs, it should be possible in the not too distant future, and using only realistic technological advances, to carry out virtual tours of exoplanets by means of a giant telescope. The financial support required would certainly exceed the cost of the trip to Mars, but the scientific and cultural adventure would be so much richer and more fruitful. In order to scrutinise the surface details of an exoplanet a hundred light years away, that is, more than a million times further than Mars, a whole network of telescopes would

have to be set up in space, covering a region as extensive as the Solar System itself. This is quite possible. Astronauts in Earth orbit would deploy several hundred mosaic mirrors, each a hundred metres across, and then send them out to the four corners of the Solar System. A few years later, pushed along by the gentle pressure of the solar wind, continually linked together by laser beams and atomic clocks, these mirrors would work together to observe the exoplanets. Their individual images would be received and pieced together on Earth using powerful computers. Such giant systems of interconnected telescopes already exist on

our planet. As large as continents, astronomers refer to them as interferometers. The new virtual hypertelescope dispersed across the Solar System would have an effective diameter of a billion kilometres and would be able to make out details the size of a human being on the surface of these distant worlds.

Hence, if there were mountains, valleys, desert plains, tides and clouds on these remote planets, they could be perfectly monitored from our own planet. As the seasons went by, we could watch this infinity of extraterrestrial landscapes gradually changing, at the whim of strange and mysterious laws. Once achieved in this idealistic way through travel without motion, the space odyssey would have attained its ultimate and most beautiful aim, hitherto considered to be utopian: to transport humanity, the whole of humanity, across the entire Universe; to traverse infinite space, and discover the stars, their planets and their moons, a multitude of other worlds.

Appendixes

CONTENTS

Yuri Gagarin

John Glenn

Gordon Cooper

Aleksei Leonov

John Young

Frank Borman

James Lovell

1961 to 2001: 397 men and women who have flown in orbit

By the beginning of 2001, just forty years after Yuri Gagarin's inaugural flight, four hundred men and women had flown in weightless conditions around the Earth. Twenty-four of them had even orbited the Moon. The table below does not list all manned flights contributing to the conquest of space since many astronauts and cosmonauts flew several times and aboard different spacecraft. Only the astronaut or cosmonaut's very first flight is recorded here, in chronological order of launch. The last column gives the number of space flights made by the astronaut. (Courtesy of Christian Lardier, Air et Cosmos.)

N°	NAME	NATIONALITY	SEX	FLIGHTS	N°	NAME	NATIONALITY	SEX	FLIGHTS
1	Yuri Gagarin	USSR 1	M	1	28	Richard Gordon	USA 17	M	2
2	Gherman Titov	USSR 2	M	1	29	Buzz Aldrin	USA 18	M	2
3	John Glenn	USA 1	M	1	30	Donn Eisele	USA 19	M	1
4	Scott Carpenter	USA 2	M	1	31	Walter Cunningham	USA 20	M	1
5	Adrian Nikolayev	USSR 3	M	2	32	Georgi Beregovoi	USSR 12	M	1
6	Pavel Popovitch	USSR 4	M	2	33	William Anders	USA 21	M	1
7	Walter Schirra	USA 3	M	3	34	Vladimir Chatalov	USSR 13	M	3
8	Gordon Cooper	USA 4	M	2	35	Boris Volynov	USSR 14	M	2
9	Valeri Bykovsky	USSR 5	M	3	36	Aleksei Yeliseyev	USSR 15	M	3
10	Valentina Tereshkova	USSR 6	F	1	37	Yevgeni Khrunov	USSR 16	M	1
11	Vladimir Komarov	USSR 7	M	2	38	Russel Schweikart	USA 22	M	1
12	Konstantin Feoktistov	USSR 8	M	1	39	Georgi Shonin	USSR 17	M	1
13	Boris Yegorov	USSR 9	M	1	40	Valeri Kubasov	USSR 18	M	3
14	Pavel Belayev	USSR 10	M	1	41	Anatoli Filipchenko	USSR 19	M	2
15	Aleksei Leonov	USSR 11	M	2	42	Vladislav Volkov	USSR 20	M	2
16	Virgil Grissom	USA 5	M	1*	43	Viktor Gorbatko	USSR 21	M	3
17	John Young	USA 6	M	6	44	Alan Bean	USA 23	M	2
18	John McDivitt	USA 7	M	2	45	John Swigert	USA 24	M	1
19	Edward White	USA 8	M	1	46	Fred Haise	USA 25	M	1
20	Charles Conrad	USA 9	M	4	47	Vitali Sevastyanov	USSR 22	M	2
21	Frank Borman	USA 10	M	2	48	Alan Shepard	USA 26	M	1*
22	James Lovell	USA 11	M	4	49	Stuart Roosa	USA 27	M	1
23	Thomas Stafford	USA 12	M	4	50	Edgar Mitchell	USA 28	M	1
24	Neil Armstrong	USA 13	M	2	51	Nikolai Rukavishnikov	USSR 23	M	3
25	David Scott	USA 14	M	3	52	Georgi Dobrovolsky	USSR 24	M	1
26	Eugene Cernan	USA 15	M	3	53	Viktor Patsayev	USSR 25	M	1
27	Michael Collins	USA 16	M	2	54	Alfred Worden	USA 29	M	1

Neil Armstrong

Michael Collins

Buzz Aldrin

James Irwin

Thomas Mattingly

Ronald Evans

Harrison Schmitt

N°	NAME	NATIONALITY	SEX	FLIGHTS	N°	NAME	NATIONALITY	SEX	FLIGHTS
55	James Irwin	USA 30	M	1	87	Vladimir Remek	Czechoslovakia	M	1
56	Thomas Mattingly	USA 31	M	3	88	Aleksandr Ivanchenkov	USSR 44	M	2
57	Charles Duke	USA 32	M	1	89	Miroslaw Hermaszewski	Poland	M	1
58	Ronald Evans	USA 33	M	1	90	Sigmund Jaehn	East Germany	M	1
59	Harrison Schmitt	USA 34	M	1	91	Vladimir Lyakhov	USSR 45	M	3
60	Joseph Kerwin	USA 35	M	1	92	Georgi Ivanov	Bulgaria	M	1
61	Paul Weitz	USA 36	M	2	93	Leonid Popov	USSR 46	M	3
62	Owen Garriott	USA 37	M	2	94	Bertalan Farkas	Hungary	M	1
63	Jack Lousma	USA 38	M	2	95	Yuri Malyshev	USSR 47	M	2
64	Vasili Lazarev	USSR 26	M	1*	96	Pham Tuan	Vietnam	M	1
65	Oleg Makarov	USSR 27	M	3*	97	Arnaldo Tamayo-Mendez	Cuba	M	1
66	Gerald Carr	USA 39	M	1	98	Leonid Kizim	USSR 48	M	3
67	Edward Gibson	USA 40	M	1	99	Gennadi Strekalov	USSR 49	M	5**
68	William Pogue	USA 41	M	1	100	Viktor Savinykh	USSR 50	M	3
69	Pyotr Klimuk	USSR 28	M	3	101	Jugderdemidiin Gurragcha	Mongolia	M	1
70	Valentin Lebedev	USSR 29	M	2	102	Robert Crippen	USA 44	M	4
71	Yuri Artyukhin	USSR 30	M	1	103	Dumitru Prunariu	Romania	M	1
72	Gennadi Sarafanov	USSR 31	M	1	104	Joe Engle	USA 45	M	2*
73	Lev Demin	USSR 32	M	1	105	Richard Truly	USA 46	M	2
74	Aleksei Gubarev	USSR 33	M	2	106	Charles Fullerton	USA 47	M	2
75	Georgi Grechko	USSR 34	M	3	107	Anatoli Berezovoi	USSR 51	M	1
76	Vance Brand	USA 42	M	4	108	Jean-Loup Chrétien	France	M	3
77	Donald Slayton	USA 43	M	1	109	Henry Hartsfield	USA 48	M	3
78	Vitali Zholobov	USSR 35	M	1	110	Aleksandr Serebrov	USSR 52	M	4
79	Vladimir Aksyonov	USSR 36	M	2	111	Svetlana Savitskaya	USSR 53	F	2
80	Vyacheslav Zudov	USSR 37	M	1	112	Robert Overmeyer	USA 49	M	2
81	Valeri Rozhdestvensky	USSR 38	M	1	113	Joseph Allen	USA 50	M	2
82	Yuri Glazkov	USSR 39	M	1	114	William Lenoir	USA 51	M	1
83	Vladimir Kovalyonok	USSR 40	M	2	115	Karol Bobko	USA 52	M	3
84	Valeri Ryumin	USSR 41	M	4	116	Donald Peterson	USA 53	M	1
85	Yuri Romanenko	USSR 42	M	3	117	Story Musgrave	USA 54	M	6
86	Vladimir Dzhanibekov	USSR 43	M	5	118	Vladimir Titov	USSR 54	M	4**

Gerald Carr

Edward Gibson

William Pogue

Vance Brand

Jack Lousma

Vladimir Titov

Sally Ride

N°	NAME	NATIONALITY	SEX	FLIGHTS	N°	NAME	NATIONALITY	SEX	FLIGHTS
119	Rick Hauck	USA 55	M	3	155	Anna Fischer	USA 84	F	1
120	John Fabian	USA 56	M	2	156	Loren Shriver	USA 85	M	3
121	Sally Ride	USA 57	F	2	157	Ellison Onizuka	USA 86	M	1**
122	Norman Thagard	USA 58	M	5	158	James Buchli	USA 87	M	4
123	Aleksandr Aleksandrov	USSR 55	M	2	159	Gary Payton	USA 88	M	1
124	Daniel Brandenstein	USA 59	M	4	160	Donald Williams	USA 89	M	2
125	Dale Gardner	USA 60	M	2	161	Jeffrey Hoffman	USA 90	M	5
126	Guion Bluford	USA 61	M	4	162	David Griggs	USA 91	M	1
127	William Thornton	USA 62	M	2	163	Margaret Seddon	USA 92	F	3
128	Brewster Shaw	USA 63	M	3	164	Jack Garn	USA 93	M	1
129	Robert Parker	USA 64	M	2	165	Frederick Gregory	USA 94	M	3
130	Byron Lichternberg	USA 65	M	2	166	Don Lind	USA 95	M	1
131	Ulf Merbold	Germany	M	3	167	Taylor Wang	USA 96	M	1
132	Robert Gibson	USA 66	M	5	168	Lodewijk Vandenberg	USA 97	M	1
133	Bruce McCandless	USA 67	M	2	169	John Creighton	USA 98	M	3
134	Ronald McNair	USA 68	M	1**	170	Shannon Lucid	USA 99	F	5
135	Robert Stewart	USA 69	M	2	171	Steven Nagel	USA 100	M	4
136	Vladimir Solovyov	USSR 56	M	2	172	Patrick Baudry	France	M	1
137	Oleg Atkov	USSR 57	M	1	173	Sultan Al Saoud	Saudi Arabia	M	1
138	Rakesh Sharma	India	M	1	174	Roy Bridges	USA 101	M	1
139	Francis Scobee	USA 70	M	1**	175	Anthony England	USA 102	M	1
140	George Nelson	USA 71	M	3	176	Karl Henize	USA 103	M	1
141	Terry Hart	USA 72	M	1	177	Loren Acton	USA 104	M	1
142	James van Hoften	USA 73	M	2	178	John Bartoe	USA 105	M	1
143	Igor Volk	USSR 58	M	1	179	Richard Covey	USA 106	M	4
144	Michael Coats	USA 74	M	3	180	John Lounge	USA 107	M	3
145	Judith Resnik	USA 75	F	1**	181	William Fischer	USA 108	M	1
146	Richard Mullane	USA 76	M	3	182	Vladimir Vasyutin	USSR 59	M	1
147	Steven Hawley	USA 77	M	4	183	Aleksandr Volkov	USSR 60	M	3
148	Charles Walker	USA 78	M	3	184	Ronald Grabe	USA 109	M	4
149	John McBride	USA 79	M	1	185	David Hilmers	USA 110	M	4
150	David Leestma	USA 80	M	3	186	William Pailes	USA 111	F	1
151	Kathryn Sullivan	USA 81	F	3	187	Bonnie Dunbar	USA 112	M	5
152	Paul Scully-Power	USA 82	M	1	188	Wubbo Ockels	Holland	M	1
153	Marc Garneau	Canada	M	2	189	Ernst Messerschmid	Germany	M	1
154	David Walker	USA 83	M	4	190	Reinhard Furrer	Germany	M	1

Bruce McCandless

Anna Fischer

Shannon Lucid

Patrick Baudry

Wubbo Ockels

Reinhard Furrer

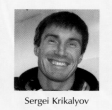
Sergei Krikalyov

N°	NAME	NATIONALITY	SEX	FLIGHTS	N°	NAME	NATIONALITY	SEX	FLIGHTS
191	Bryan O'Connor	USA 113	M	2	227	John Casper	USA 137	M	4
192	Sherwood Spring	USA 114	M	1	228	Pierre Thuot	USA 138	M	3
193	Jerry Ross	USA 115	M	5	229	Gennadi Manakov	USSR 69	M	2
194	Mary Cleave	USA 116	F	2	230	Robert Cabana	USA 139	M	3
195	Rodolfo Neri Vela	Mexico	M	1	231	Thomas Akers	USA 140	M	4
196	Charles Bolden	USA 117	M	4	232	Bruce Melnick	USA 141	M	2
197	Franklin Chang-Diaz	USA 118	M	6	233	Frank Culbertson	USA 142	M	2
198	Robert Cenker	USA 119	M	1	234	Charles Gemar	USA 143	M	3
199	William Nelson	USA 120	M	1	235	Carl Meade	USA 144	M	3
200	Aleksandr Laveykin	USSR 61	M	1	236	Ronald Parise	USA 145	M	1
201	Aleksandr Viktorenko	USSR 62	M	4	237	Samuel Durrance	USA 146	M	1
202	Mohammed Faris	Syria	M	1	238	Viktor Afanasyev	USSR 70	M	3
203	Musa Manarov	USSR 63	M	2	239	Tohiro Akiyama	Japan	M	1
204	Anatoli Levchenko	USSR 64	M	1	240	Kenneth Cameron	USA 147	M	3
205	Anatoli Solovyov	USSR 65	M	5	241	Jerome Apt	USA 148	M	4
206	Aleksandr Aleksandrov	Bulgaria	M	1	242	Linda Godwin	USA 149	F	3
207	Valeri Polyakov	USSR 66	M	2	243	Lloyd Hammond	USA 150	M	2
208	Abdul Ahad Mohmand	Afghanistan	M	1	244	Richard Hieb	USA 151	M	3
209	Sergei Krikalyov	USSR 67	M	3	245	Gregory Harbaugh	USA 152	M	4
210	Guy Gardner	USA 121	M	2	246	Donald McMonagle	USA 153	M	3
211	William Shepherd	USA 122	M	3	247	Charles Veach	USA 154	M	2
212	John Blaha	USA 123	M	5	248	Anatoli Artsebarsky	USSR 71	M	1
213	Robert Springer	USA 124	M	2	249	Helen Sharman	UK	F	1
214	James Bagian	USA 125	M	2	250	Sidney Gutierriez	USA 155	M	2
215	Mark Lee	USA 126	M	4	251	Tamara Jernigan	USA 156	F	4
216	Richard Richards	USA 127	M	4	252	Andrew Gaffney	USA 157	M	1
217	James Adamson	USA 128	M	2	253	Millie Hughes-Fulford	USA 158	F	1
218	Mark Brown	USA 129	M	2	254	Michael Baker	USA 159	M	4
219	Michael McCulley	USA 130	M	1	255	Kenneth Reightler	USA 160	M	2
220	Ellen Shulman-Baker	USA 131	F	3	256	Toktar Aubakirov	Kazakhstan	M	1
221	Kathryn Thornton	USA 132	F	4	257	Franz Viehboeck	Austria	M	1
222	Manley Carter	USA 133	M	1	258	Terence Henricks	USA 161	M	4
223	James Wetherbee	USA 134	M	4	259	James Voss	USA 162	M	3
224	David Low	USA 135	M	3	260	Mario Runco	USA 163	M	3
225	Marsha Ivins	USA 136	F	4	261	Thomas Hennen	USA 164	M	1
226	Aleksandr Balandin	USSR 68	M	1	262	Stephen Oswald	USA 165	M	3

Kathryn Thornton

James Wetherbee

Roberta Bondar

Steven MacLean

Hans Schlegel

Chiaki Mukai

Jerry Linenger

N°	NAME	NATIONALITY	SEX	FLIGHTS	N°	NAME	NATIONALITY	SEX	FLIGHTS
263	William Readdy	USA 166	M	3	299	James Newman	USA 187	M	2
264	Roberta Bondar	Canada	F	1	300	Carl Walz	USA 188	M	3
265	Aleksandr Kaleri	Russia 72	M	2	301	Richard Searfoss	USA 189	M	2
266	Klaus-Dietrich Flade	Germany	M	1	302	William McArthur	USA 190	M	2
267	Brian Duffy	USA 167	M	3	303	David Wolf	USA 191	M	2
268	Michael Foale	USA 168	M	4	304	Martin Fettman	USA 192	M	1
269	Dirk Frimout	Belgium	M	1	305	Yuri Usachyov	Russia 76	M	2
270	Kevin Chilton	USA 169	M	3	306	Ronald Sega	USA 193	M	2
271	Kenneth Bowersox	USA 170	M	4	307	Thomas Jones	USA 194	M	3
272	Lawrence DeLucas	USA 171	M	1	308	Yuri Malenchenko	Russia 77	M	1
273	Eugene Trinh	USA 172	M	1	309	Talgat Musabayev	Russia 78	M	2
274	Sergei Avdeyev	Russia 73	M	3	310	James Halsell	USA 195	M	4
275	Michel Tognini	France	M	2	311	Leroy Chaio	USA 196	M	2
276	Andrew Allen	USA 173	M	3	312	Donald Thomas	USA 197	M	4
277	Franco Malerba	Italy	M	1	313	Chiaki Mukai	Japan	F	1
278	Claude Nicollier	Switzerland	M	3	314	Jerry Linenger	USA 198	M	2
279	Curtis Brown	USA 174	M	4	315	Terrence Wilcutt	USA 199	M	3
280	Jan Davis	USA 175	F	3	316	Steven Smith	USA 200	M	2
281	Mae Jemison	USA 176	F	1	317	Elena Kondakova	Russia 79	F	2
282	Mamoru Mohri	Japan	M	1	318	Joseph Tanner	USA 201	M	1
283	Steven MacLean	Canada	M	1	319	Jean-François Clervoy	France	M	3
284	Michael Clifford	USA 177	M	3	320	Scott Parazynski	USA 202	M	2
285	Susan Helms	USA 178	F	3	321	Eileen Collins	USA 203	F	2
286	Aleksandr Polishchuk	Russia 74	M	2	322	William Gregory	USA 204	M	1
287	Kenneth Cockrell	USA 179	M	3	323	John Grunsfeld	USA 205	M	2
288	Ellen Ochoa	USA 180	F	2	324	Wendy Lawrence	USA 206	F	3
289	Charles Precourt	USA 181	M	4	325	Vladimir Dezhurov	Russia 80	M	1
290	Bernard Harris	USA 182	M	2	326	Nikolai Budarin	Russia 81	M	2
291	Hans Schlegel	Germany	M	1	327	Kevin Kregel	USA 207	M	4
292	Ulrich Walter	Germany	M	1	328	Mary Weber	USA 208	F	1
293	Nancy Sherlock-Currie	USA 183	F	2	329	Yuri Gidzenko	Russia 82	M	1
294	Peter Wisoff	USA 184	M	3	330	Thomas Reiter	Germany	M	1
295	Janice Voss	USA 185	F	4	331	Michael Gernhardt	USA 209	M	4
296	Vasili Tsibliyev	Russia 75	M	2	332	Kent Rominger	USA 210	M	3
297	Jean-Pierre Haigneré	France	M	2	333	Catherine Coleman	USA 211	F	1
298	Daniel Bursch	USA 186	M	3	334	Michael Lopez-Alegria	USA 212	M	1

Yuri Gidzenko

Koichi Wakata

Maurizio Cheli

Jean-Jacques Favier

Bjarni Tryggvason

Dominic Gorie

Frederick Sturckow

N°	NAME	NATIONALITY	SEX	FLIGHTS	N°	NAME	NATIONALITY	SEX	FLIGHTS
335	Albert Sacco	USA 213	M	1	371	James Reilly	USA 234	M	1
336	Fred Leslie	USA 214	M	1	372	Saliszan Sharipov	Russia 87	M	1
337	Chris Hadfield	Canada	M	1	373	Léopold Eyharts	France	M	1
338	Brent Jett	USA 215	M	2	374	Scott Altman	USA 235	M	1
339	Daniel Barry	USA 216	M	2	375	Kathryn Hire	USA 236	F	1
340	Winston Scott	USA 217	M	1	376	Dafydd Williams	Canada	M	1
341	Koichi Wakata	Japan	M	1	377	Jay Buckey	USA 237	M	1
342	Yuri Onufrienko	Russia 83	M	1	378	James Pawelczyk	USA 238	M	1
343	Scott Horowitz	USA 218	M	2	379	Dominic Gorie	USA 239	M	1
344	Umberto Guidoni	Italy	M	1	380	Janet Kavandi	USA 240	F	1
345	Maurizio Cheli	Italy	M	1	381	Gennadi Padalka	Russia 88	M	1
346	Andrew Thomas	USA 219	M	2	382	Yuri Baturin	Russia 89	M	1
347	Richard Linnehan	USA 220	M	1	383	Pedro Duque	Spain	M	1
348	Charles Brady	USA 221	M	1	384	Frederick Sturckow	USA 241	M	1
349	Jean-Jacques Favier	France	M	1	385	Ivan Bella	Slovakia	M	1
350	Robert Thirsk	Canada	M	1	386	Rick Husband	USA 242	M	1
351	Valeri Korzun	Russia 84	M	1	387	Julie Payette	Canada	F	1
352	Claudie André-Deshays	France	F	1	388	Valeri Tokarev	Russia 90	M	1
353	Aleksandr Lazutkin	Russia 85	M	1	389	Jeffrey Ashby	USA 243	M	1
354	Reinhold Ewald	Germany	M	1	390	Scott Kelly	USA 244	M	1
355	Susan Still	USA 222	F	2	391	Gerhard Thiele	Germany	M	1
356	Roger Crouch	USA 223	M	2	392	Sergei Zalyotin	Russia 91	M	1
357	Gregory Linteris	USA 224	M	2	393	Jeffrey Williams	USA 245	M	1
358	Carlos Noriega	USA 225	M	1	394	Daniel Burbank	USA 246	M	1
359	Edward Lu	USA 226	M	1	395	Richard Mastracchio	USA 247	M	1
360	Pavel Vinogradov	Russia 86	M	1	396	Boris Morukov	Russia 92	M	1
361	Robert Curbeam	USA 227	M	1	397	Pamela Melroy	USA 248	F	1
362	Stephen Robinson	USA 228	M	1					
363	Bjarni Tryggvason	Canada	M	1					
364	Michael Bloomfield	USA 229	M	1					
365	Steven Lindsey	USA 230	M	1					
366	Kalpana Chawla	USA 231	F	1					
367	Takao Doi	Japan	M	1					
368	Leonid Kadenyuk	Ukraine	M	1					
369	Joe Edwards	USA 232	M	1					
370	Michael Anderson	USA 233	M	1					

* Also made a suborbital flight.

** Also had one launch failure.

Source : Christian Lardier – Air et Cosmos

NOTES

[1] Michael Smith, Gregory Jarvis and Christa McAuliffe, who were aboard *Challenger* in 1986, never flew in space.

[2] Aubakirov flew as a citizen of the Republic of Kazakhstan whilst Musabayev (also born in Kazakhstan) flew as a Russian, first as flight engineer, then flight commander.

[3] Sevenx15 pilots also made thirteen suborbital flights, receiving their astronaut's wings.

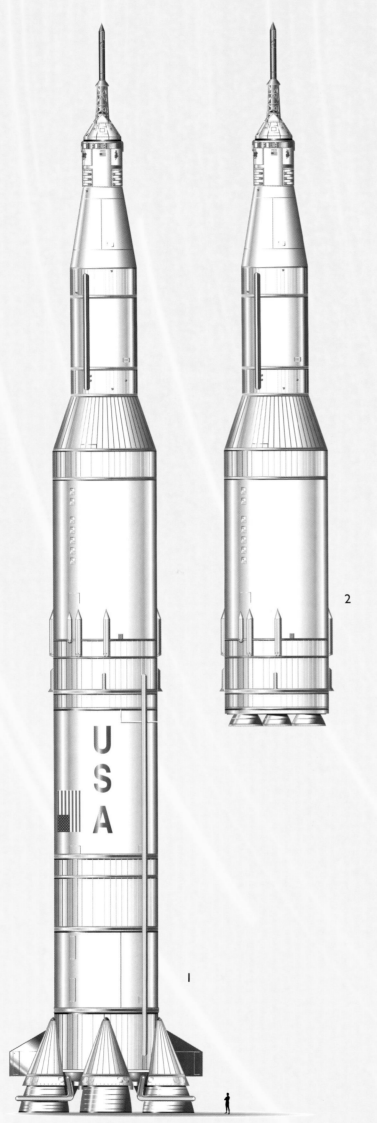

3

7

10

The Apollo programme: A typical mission

Saturn V is one of the most powerful launch vehicles ever built. Its power and cargo capacity were only slightly exceeded by the Soviet rocket Energia, tested in 1987 and 1988 but never actually used. The spearhead of the Apollo programme flew for the first time in 1967. It measured 110 metres high and 10 metres across at the base. Empty, it weighed in at 190 tonnes, whilst at blast-off, its tanks filled with fuel and oxidising agent, the rocket reached 2 950 tonnes. The thrust delivered by the engines in the first propulsion stage (1), burning a mixture of kerosene and liquid oxygen, approached 3 500 tonnes, roughly equivalent to 150 million horsepower. The second stage (2) produced 450 tonnes of thrust, whilst the third and last stage (3) added a further 90 tonnes. The two last stages burnt a mixture of hydrogen and liquid oxygen, propelling the whole of the space train towards the Moon at 39 600 km/hr. To gain space, NASA's engineers packed the lunar module upside-down in the rocket fairing. During the journey to the Moon (7), the command and service module (CSM) and lunar module (LM) were released from the aerodynamic fairing of the third stage in order to redock together in the correct configuration (10). The three components of the spacecraft – command module, service module and lunar module – weighed over forty tonnes in all.

29

22

35

The LM which landed on the Moon (22) weighed about fifteen tonnes. However, in the Moon's much smaller gravitational field, six times weaker than the Earth's, it weighed the same as a mass of only 2.5 tonnes on Earth. This is one reason why much less energy is required to escape from the lunar surface. Another is that there is no atmosphere to cause friction, air braking and heating, as happens on Earth. Furthermore, only the upper part of the LM actually took off, the so-called ascent stage, using a small engine with just 1.5 tonnes of thrust. The descent stage was left forever at the lunar landing site. Meanwhile, the ascent stage redocked with the CSM in lunar orbit and set off back to Earth (29). When the re-entry capsule finally swung down through the Earth's atmosphere on the end of its three huge parachutes (35), the remains of the Apollo spacecraft weighed a mere five tonnes, compared with its original lift-off mass of 2950 tonnes. This is the tribute that must be paid to gravity if human beings are to free themselves from the Earth's grasp and travel freely through space.

After departure (1), the Saturn V rocket jettisoned its first and second stages (2,3), and went into orbit around the Earth (4). After one revolution, it began the long space crossing (5) during which the astronauts would turn the LM around (7,8,9,10) and jettison the third stage (13). The little space train was now ready to acquire its orbit around the Moon (11,12,14,15,16). The LM then separated from the CSM (17,18) and landed on the Moon (19), whilst the CSM continued in its lunar orbit (20,21). When it left the Moon's surface, the LM ascent stage returned to the orbiting CSM and redocked (23,24,25). The LM ascent stage was then jettisoned to crash onto the Moon (26), whilst the three astronauts returned to Earth aboard the

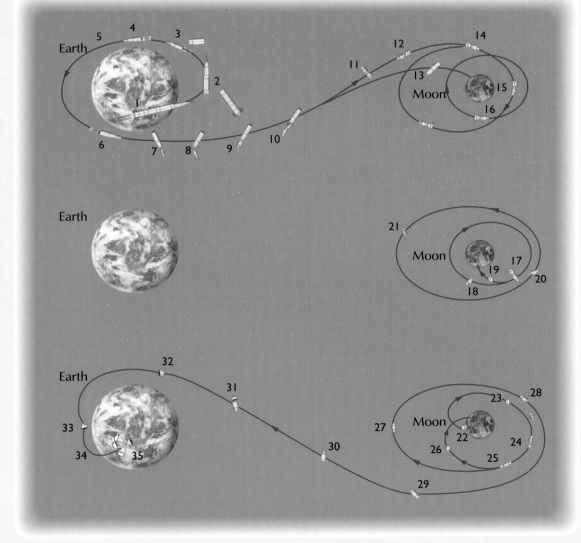

CSM (27,28,29,30). Finally, the command module was detached from the service module (31,32), entering the upper layers of the Earth's atmosphere (33,34) and deploying its three great parachutes above the Pacific Ocean (35). This is how, between 1968 and 1972, twenty-four men returned safe and sound from the Moon.

American and Soviet missions to the Moon

About twenty Soviet and American missions successfully reached the Moon. The most important missions are represented here, manned for the Americans and unmanned for the Soviets. The Apollo programme allowed twelve men to walk on the Moon and bring back a total of 385 kg of lunar dust and rock samples.

The *Lunakhod* vehicles carried to the Moon by Luna 17 and Luna 21 travelled ten kilometres and thirty-seven kilometres, respectively, across its dusty plains. Rockets carried by Luna 16, Luna 20 and Luna 24 allowed the Soviets to bring back lunar samples to Earth without the need for human presence.

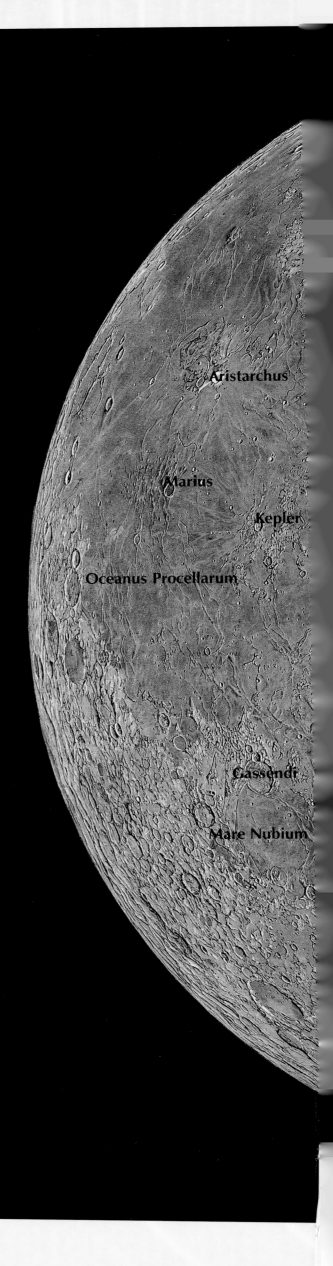

Aristarchus

Marius

Kepler

Oceanus Procellarum

Cassendi

Mare Nubium

Soviets and Americans never trod the lunar surface together but they did collaborate for the Apollo Soyuz Test Project (ASTP). This first common space mission aimed to test the use of cooperation for crew rescue. The photograph shows American astronaut Thomas P. Stafford training with Soviet cosmonaut Aleksei A. Leonov whose head appears through the hatch of the Soyuz mock-up. On 17 July 1975, an Apollo spacecraft and a Soyuz spacecraft succeeded in docking for a period of forty-seven hours at an altitude of 225 kilometres.

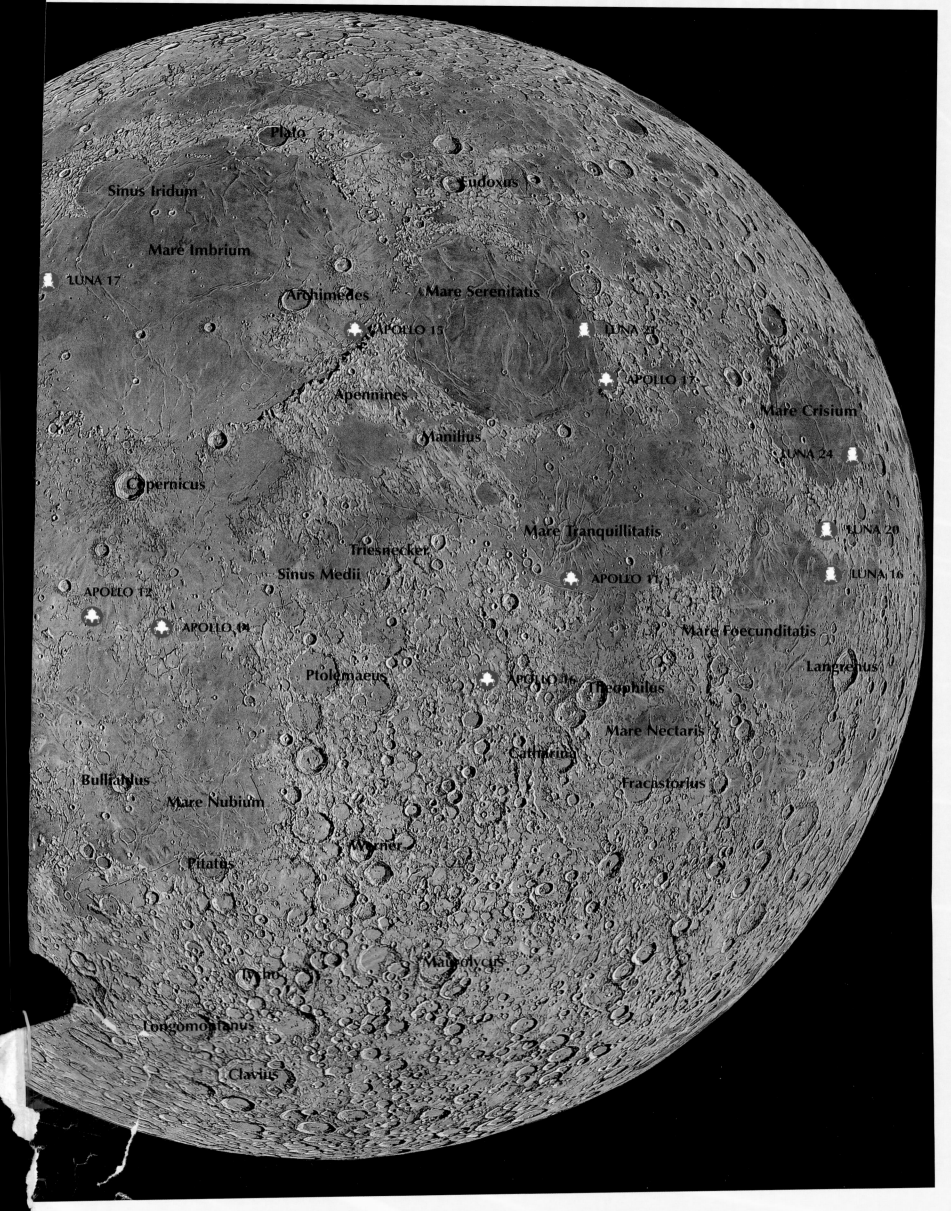

The Mir space station

① MIR. The first module of the space station was Mir, situated at the heart of the construction and affording a living space of over 100 m³. It was sent into orbit at an altitude of 350 kilometres on 19 February 1986.

② KVANT. Another module was docked onto Mir in 1987. *Kvant* was equipped with several X-ray telescopes capable of detecting very high energy radiation produced by hot stars.

③ KVANT 2. With the addition of *Kvant 2* in 1989, the space station became a much more comfortable place to live. It contained a water supply and was equipped with showers as well as an EVA hatch.

④ KRISTALL. In 1990 a further module docked onto the space station. *Kristall* was twelve metres long and weighed almost 20 tonnes. It was equipped with solar arrays and supplied energy to the space station.

⑤ SPEKTR. Docked onto the Russian station in 1995, *Spektr* was accidentally knocked by the Progress resupply vessel and depressurised, wherupon it had to be abandoned. It had been devoted to Earth and atmospheric observations.

⑥ PRIRODA. The construction of the Mir space station was completed in 1996 when the *Priroda* module was attached. During fifteen years of service, more than a hundred cosmonauts and astronauts were welcomed aboard.

⑦ SHUTTLE DOCKING MODULE. Between 1995 and 1999, nine American shuttle missions were organised in collaboration with the Russians in order to prepare for the future International Space Station. This module was specially designed to allow the shuttle to dock onto Mir.

⑧ PROGRESS RESUPPLY VESSEL. Regular supplies had to be delivered and this was achieved by the Progress spacecraft. In addition, Mir was slowed down as it passed through the upper atmosphere. Progress was used to push it back onto its nominal orbit.

⑨ SOYUZ SPACECRAFT. Two or three cosmonauts could be ferried out to Mir by the Soyuz spacecraft. One Soyuz was permanently docked onto the space station, ready for immediate return to Earth in the event of accident.

⑩ SOFORA BOOM ARM. Mir was equipped with a huge mast whose main purpose was active stabilisation of the space station using dedicated thrusters. It also served to install antennae and scientific equipment such as micrometeorite detectors.

⑪ SOLAR PANELS. Mir required a large amount of energy and this was generously supplied by the Sun. The large solar arrays carried by the space station provided a power of 34 kW.

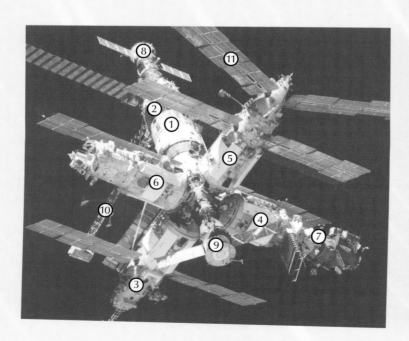

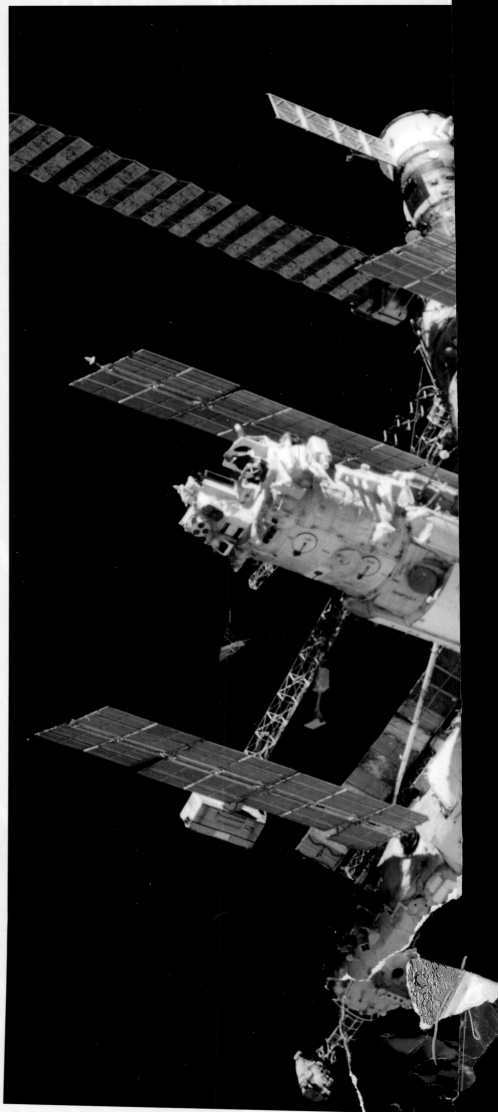

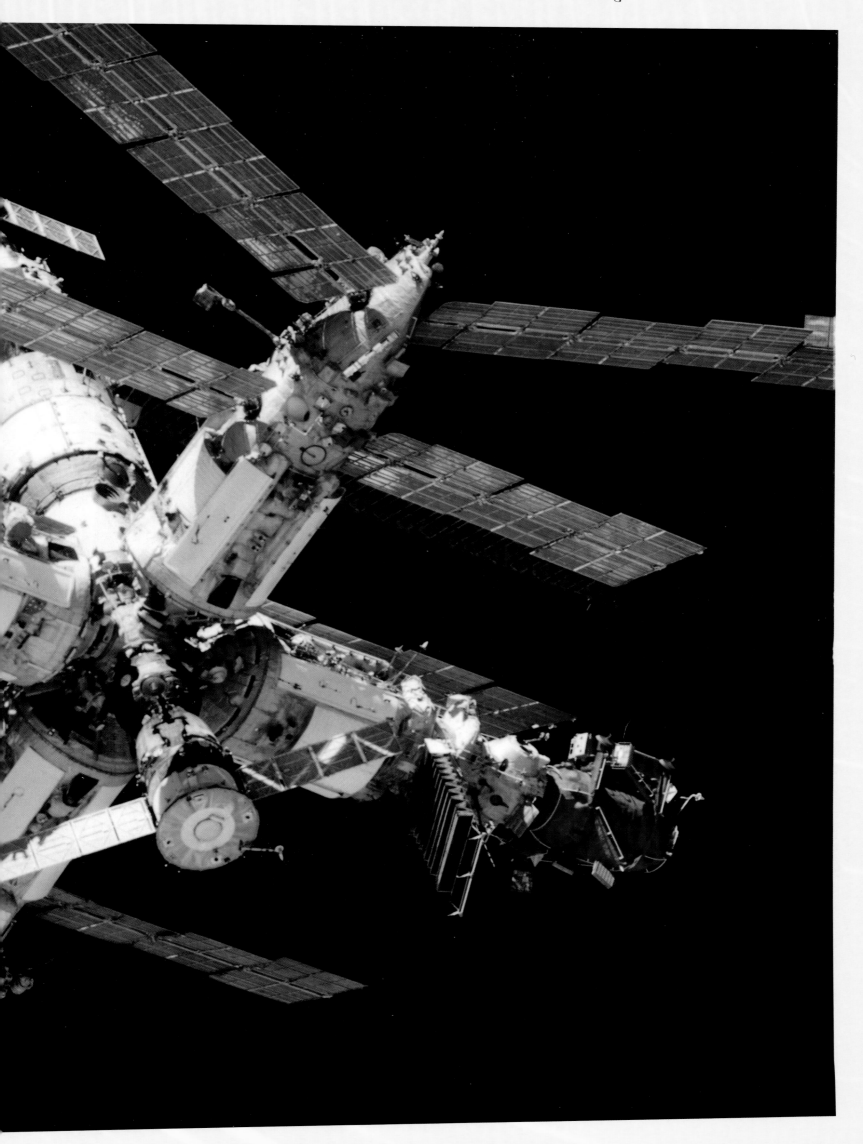

Chronology of the first 100 flights of the US sp

FLIGHT	NUMBER	LAUNCH DATE	MISSION	SHUTTLE
1	STS 1	12 April 1981	Inaugural flight of space shuttle	*Columbia*
2	STS 2	12 November 1981	OSTA 1 experiment	*Columbia*
3	STS 3	22 March 1982	OSS 1 experiment	*Columbia*
4	STS 4	27 June 1982	US Department of Defense	*Columbia*
5	STS 5	11 November 1982	Anik C3 and SBS C satellites	*Columbia*
6	STS 6	4 April 1983	TDRS 1 satellite	*Challenger*
7	STS 7	18 June 1983	Anik C2 and SBS C satellites	*Challenger*
8	STS 8	30 August 1983	Insat 1B satellite	*Challenger*
9	STS 9	28 November 1983	Spacelab 1	*Columbia*
10	STS 41B	3 February 1984	Westar VI and Palapa BI satellites	*Challenger*
11	STS 41C	6 April 1984	LDEF experiment	*Challenger*
12	STS 41D	30 August 1984	SBS D, Syncom IV 2 and Telstar satellites	*Discovery*
13	STS 41G	5 October 1984	ERBS satellite and OSTA 3 experiment	*Challenger*
14	STS 51A	8 November 1984	Telesat H and Syncom IV 1 satellites	*Discovery*
15	STS 51C	24 January 1985	US Department of Defense	*Discovery*
16	STS 51D	12 April 1985	Telesat 1 and Syncom IV 3 satellites	*Discovery*
17	STS 51B	29 April 1985	Spacelab 3	*Challenger*
18	STS 51G	17 June 1985	Morelos, Arabsat A and Telstar satellites	*Discovery*
19	STS 51F	29 July 1985	Spacelab 2	*Challenger*
20	STS 51I	27 August 1985	ASC 1, Aussat 1 and Syncom IV 4 satellites	*Discovery*
21	STS 51J	3 October 1985	US Department of Defense	*Atlantis*
22	STS 61A	30 October 1985	Spacelab D1 and GLOMR satellite	*Challenger*
23	STS 61B	26 November 1985	Morelos B, Aussat 2 and Satcom KU 2 satellites	*Atlantis*
24	STS 61C	12 January 1986	Satcom KU 1	*Columbia*
25	STS 51L	28 January 1986	TDRS B and Spartan 203 satellites	*Challenger*
26	STS 26	29 September 1988	TDRS C satellite	*Discovery*
27	STS 27	2 December 1988	US Department of Defense	*Atlantis*
28	STS 29	13 March 1989	TDRS D satellite	*Discovery*
29	STS 30	4 May 1989	Launch of Venus probe Magellan	*Atlantis*
30	STS 28	8 August 1989	US Department of Defense	*Columbia*
31	STS 34	18 October 1989	Launch of Jupiter probe Galileo	*Atlantis*
32	STS 33	22 November 1989	US Department of Defense	*Discovery*
33	STS 32	9 January 1990	LDEF experiment and Syncom IV F5 satellite	*Columbia*
34	STS 36	28 February 1990	US Department of Defense	*Atlantis*
35	STS 31	24 April 1990	Deployment of Hubble Space Telescope	*Discovery*
36	STS 41	6 October 1990	Launch of solar probe Ulysses	*Discovery*
37	STS 38	15 November 1990	US Department of Defense	*Atlantis*
38	STS 35	2 December 1990	Astro 1 space laboratory	*Columbia*
39	STS 37	5 April 1991	Launch of Compton Gamma Ray Observatory	*Atlantis*
40	STS 39	28 April 1990	US Department of Defense	*Discovery*
41	STS 40	5 June 1991	SLS 1 spacelab	*Columbia*
42	STS 43	2 August 1991	TDRS E satellite	*Atlantis*
43	STS 48	12 September 1991	UARS satellite	*Discovery*
44	STS 44	24 November 1991	US Department of Defense DSP 16 satellite	*Atlantis*
45	STS 42	22 January 1992	IML 1 experiment	*Discovery*
46	STS 45	24 March 1992	Atlas 1 space laboratory	*Atlantis*
47	STS 49	7 May 1992	Repairs to Intelsat VI satellite	*Endeavour*
48	STS 50	25 June 1992	USML 1 experiment	*Columbia*
49	STS 46	31 July 1992	Eureca experiment and TSS 1 satellite	*Atlantis*
50	STS 47	12 September 1992	Spacelab J	*Endeavour*
51	STS 52	22 October 1992	USMP 1 experiment and Lageos II satellite	*Columbia*
52	STS 53	2 December 1992	US Department of Defense	*Discovery*
53	STS 54	13 January 1993	TDRS F satellite	*Endeavour*
54	STS 56	8 April 1993	Atlas 2 space laboratory and Spartan satellite	*Discovery*

ace shuttle (1981 to 2000)

FLIGHT	NUMBER	LAUNCH DATE	MISSION	SHUTTLE
55	STS 55	26 April 1993	Spacelab D2	*Columbia*
56	STS 57	21 June 1993	Spacehab and Eureka laboratories	*Endeavour*
57	STS 51	12 September 1993	ACTS and ORFEUS-SPAS satellites	*Discovery*
58	STS 58	18 October 1993	SLS 2 spacelab	*Columbia*
59	STS 61	2 December 1993	Repairs to Hubble Space Telescope	*Endeavour*
60	STS 60	3 February 1994	Spacehab 2 laboratory	*Discovery*
61	STS 62	4 March 1994	USMP 2 and OAST 2 experiments	*Columbia*
62	STS 59	9 April 1994	SRL 1 space laboratory	*Endeavour*
63	STS 65	8 July 1994	IML 2 space laboratory	*Columbia*
64	STS 64	9 September 1994	LITE lidar and Spartan 201 experiments	*Discovery*
65	STS 68	30 September 1994	SRL 2 space laboratory	*Endeavour*
66	STS 66	3 November 1994	Atlas 3 space laboratory, CRISTA-SPAS 1 satellite	*Atlantis*
67	STS 63	3 February 1995	Mir rendezvous, Spartan 204, Oderacs, Spacehab 3	*Discovery*
68	STS 67	2 March 1995	Astro 2 space laboratory	*Endeavour*
69	STS 71	27 June 1995	First docking with Mir (SMM 1)	*Atlantis*
70	STS 70	13 July 1995	TDRS G satellite	*Discovery*
71	STS 69	7 September 1995	WSF 2 and Spartan 201 experiments	*Endeavour*
72	STS 73	20 October 1995	USML 2 space laboratory	*Columbia*
73	STS 74	12 November 1995	Second docking with Mir (SMM 2)	*Atlantis*
74	STS 72	11 January 1996	SFU and OAST-Flyer experiments	*Endeavour*
75	STS 75	22 February 1996	TSS 1R and USMP 3 experiments	*Columbia*
76	STS 76	22 March 1996	Third docking with Mir (SMM 3)	*Atlantis*
77	STS 77	19 May 1996	Spacehab 4 laboratory, Spartan 207 experiment	*Endeavour*
78	STS 78	20 June 1996	Spacelab LMS	*Columbia*
79	STS 79	16 September 1996	Fourth docking with Mir (SMM 4)	*Atlantis*
80	STS 80	19 November 1996	ORFEUS-SPAS II and WSF 3 experiments	*Columbia*
81	STS 81	12 January 1997	Fifth docking with Mir (SMM 5)	*Atlantis*
82	STS 82	11 February 1997	Maintenance on Hubble Space Telescope	*Discovery*
83	STS 83	4 April 1997	MSL 1 space laboratory	*Columbia*
84	STS 84	15 May 1997	Sixth docking with Mir (SMM 6)	*Atlantis*
85	STS 94	1 July 1997	MSL 1R space laboratory	*Columbia*
86	STS 85	7 August 1997	CRISTA-SPAS 2 experiment	*Discovery*
87	STS 86	25 September 1997	Seventh docking with Mir (SMM 7)	*Atlantis*
88	STS 87	19 November 1997	USMP 4 and Spartan 201 experiments	*Columbia*
89	STS 89	22 January 1998	Eighth docking with Mir (SMM 8)	*Endeavour*
90	STS 90	17 April 1998	Neurolab space laboratory	*Columbia*
91	STS 91	2 June 1998	Final docking with Mir (SMM 9)	*Discovery*
92	STS 95	19 October 1998	Spacehab laboratory and Spartan 201 experiment	*Discovery*
93	STS 88	4 December 1998	ISS connecting module Unity, SAC A, Mightysat	*Endeavour*
94	STS 96	27 May 1999	Docking with ISS, Starshine satellite	*Discovery*
95	STS 93	22 July 1999	Launch of Chandra X-ray telescope	*Columbia*
96	STS 103	19 December 1999	Maintenance on Hubble Space Telescope	*Discovery*
97	STS 99	11 February 2000	SRTM radar experiment	*Endeavour*
98	STS 101	19 May 2000	Docking with ISS	*Atlantis*
99	STS 106	8 September 2000	Docking with ISS	*Atlantis*
100	STS 92	11 October 2000	Docking with ISS	*Discovery*

In the year 2000, NASA celebrated the hundredth space shuttle launch, including only one failure when *Challenger* blew up after launch, causing the loss of all seven astronauts in January 1986. Over the twenty-year period, the five American shuttles underwent many improvements to onboard computer systems, materials used to protect outer surfaces, and safety and life support systems. The shuttles *Atlantis*, *Columbia*, *Discovery* and *Endeavour* will probably continue to fly for a further twenty years. It is hoped to enhance both performance and reliability by equipping them with newly-developed cryogenic propulsion stages. (Note: STS stands for Space Transportation System.)

Chronology of French missions up to 2000

MISSION	LAUNCH DATE	COUNTRIES INVOLVED	SPATIONAUT	SPACECRAFT
PVH	24 June 1982	France/USSR	Jean-Loup Chrétien	Salyut 7
51 G	17 June 1985	France/USA	Patrick Baudry	*Discovery*
Aragatz	26 November 1988	France/USSR	Jean-Loup Chrétien	Mir
Antares	27 July 1992	France/Russia	Michel Tognini	Mir
Altair	1 July 1993	France/Russia	Jean-Pierre Haigneré	Mir
STS 66	3 November 1994	USA/Europe	Jean-François Clervoy	*Atlantis*
STS 78	20 June 1996	France/USA	Jean-Jacques Favier	*Columbia*
Cassiopeia	17 August 1996	France/Russia	Claudie André-Deshays	Mir
STS 84	15 May 1997	USA/Europe	Jean-François Clervoy	*Atlantis*/Mir
STS 86	25 September 1997	France/USA	Jean-Loup Chrétien	*Atlantis*/Mir
Pegasus	29 January 1998	France/Russia	Léopold Eyharts	Mir
Perseus	20 February 1999	France/Russia	Jean-Pierre Haigneré	Mir
STS 93	22 July 1999	France/USA	Michel Tognini	*Columbia*
STS 103	19 December 1999	USA/Europe	Jean-François Clervoy	*Discovery*

During the 1980s and 1990s, eight French spationauts flew in space in collaboration with the Soviet Union, the Russian Federation, and the United States, then in the context of the European Space Agency. In the future, French spationauts working for the ESA will live aboard the International Space Station or take part in US space shuttle missions.

■ JEAN-PIERRE HAIGNERÉ ABOARD MIR.

The ambassadors:
Trajectories followed by Pioneer 10, Pioneer 11, Voyager 1 and Voyager 2

Four space probes are currently leaving the Solar System. Despite their tremendous distance from Earth, three of them can still be monitored by NASA's Deep Space Network. The diagram shows the trajectories of the four probes in relation to the orbits of the more remote planets which they helped to study in detail for the first time. Distances from the Sun are given in billions of kilometres. Orbits and probe distances have been projected onto the plane of the ecliptic to provide a simpler view.

Pioneer 10 left Earth in 1972, followed by Pioneer 11 in 1973. The latter stopped transmitting in 1995. Voyager 1 and 2 left Earth in 1977.

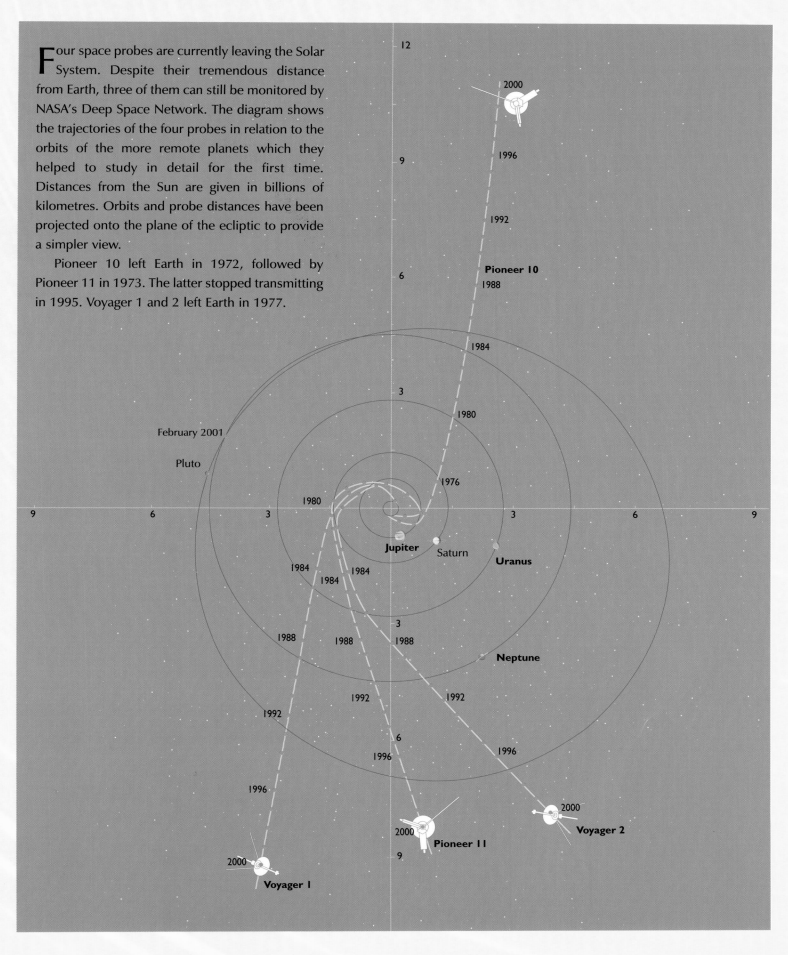

The International Space Station

When completed in 2006, the International Space Station will include a Japanese and a European module as well as Russian and American modules. The remote manipulator arm used to move modules and other equipment around in space has been designed and built by a Canadian team. The very first module of the station was *Zarya*, stationed in orbit at the end of November 1998. The connecting module *Unity* was docked in December 1998, followed by the *Zvezda* service module in July 2000. The *Destiny* laboratory module was added in February 2001.

The International Space Station hosted its first Russian–American crew (expedition one) from 2 November 2000 until 21 March 2001, when it was replaced by the expedition two crew, made up of two Americans and a Russian. Future crews will contain representatives from all participating countries. Measuring eighty metres long and weighing more than 450 tonnes, the completed space station will be able to house crews of six to seven astronauts in a living space of 1 300 cubic metres. This is comparable with the space available aboard a Boeing 747. Eight pairs of solar panels will be deployed on either side of the station along a mast 108 metres long, supplying a power of over 100 kW.

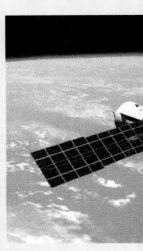

1998

2002

2004

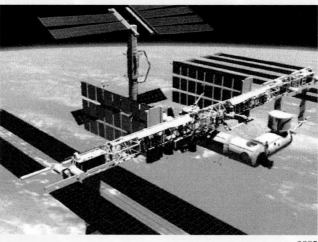

2005

1998

2000

2002

2003

2003

2003

2004

2004

2005

2006

2006

2006

INDEX

BIBLIOGRAPHY

Joseph Allen: *Aller-retour pour l'espace*, Mazarine, 1984.

Jay Apt, Michael Helfert, Justin Wilkinson: *Orbit*, National Geographic Society, 1996.

Jacques Blamont: *Vénus dévoilée*, Editions Odile Jacob, 1987.

Roger-Maurice Bonnet: *Les horizons chimériques*, Dunod, 1992.

Bryan Burrough: *Dragonfly: NASA and the crisis aboard Mir*, HarperCollins Publishers, 1998.

Jean-Loup Chrétien: *Mission Mir, journal de bord*, Michel Lafon, 1998.

Jean-Pierre Defait: *50 ans de conquête spatiale*, special issue of Ciel et Espace, 1999.

Alain Dupas: *Une autre histoire de l'espace*, Découvertes Gallimard, 1999.

Albert Einstein, Leopold Infeld: *Evolution of physics: From early concepts to relativity and quanta*, Simon and Schuster, 1938.

Charles Frankel: *La vie sur Mars*, Science ouverte, Seuil, 1999.

Olivier de Goursac: *A la conquête de Mars*, Larousse, 2000.

Serguëi Grichkov: *Guide des lanceurs spatiaux*, Tessier & Ashpool, 1992.

Kevin W. Kelley: *The Home Planet*, Macdonald, 1988.

Pierre Kohler: *Les grandes heures des conquérants de l'espace*, Librairie Perrin, 1989.

Pierre Kohler: *La dernière mission*, Calmann-Lévy, 2000.

Alexandre Koyré: *Du monde clos à l'univers infini*, Tel, Gallimard, 1973.

Christian Lardier: *L'astronautique sovietique*, Armand Colin, 1992.

Michael Light: *Full moon*, Random House, 1999.

Laurent Nottale: *La relativité dans tous ses états*, Hachette, 1998.

Nikos Prantzos: *Our cosmic future: Humanity's fate in the Universe*, Cambridge University Press, 2000.

Alan Shepard, Deke Slayton: *Moon shot. The inside story of America's race to the Moon*, Turner Publishing, 1994.

Fernand Verger: *The Cambridge Encyclopedia of Space*, Cambridge University Press, 2002.

Jacques Villain: *A la conquête de la Lune*, Larousse, 1998.

Jacques Villain: *Baïkonur, la porte des étoiles*, Armand Colin, 1994.

Le grand atlas de l'espace, Encyclopedia Universalis, 1989.

Websites

Official NASA website: http://www.nasa.gov

Official ESA website: http://www.esa.int

Official website for the French space agency (CNES): http://www.cnes.fr

Science at the European Space Agency: http://sci.esa.int

All scientific space missions: http://spacescience.nasa.gov/missions/index.htm

NASA human spaceflight: http://spaceflight.nasa.gov/index-m.html

NASA's Mars exploration programme: http://spaceflight.nasa.gov/mars/index.html

International Space Station: http://spaceflight.nasa.gov/station/index.html

Mark Wade's encyclopedia of astronautics: http://www.friends-partners.ru/partners/mwade/spaceflt.htm

European launcher programme: http://www.esa.int/launchers

Apollo-Saturn launcher programme: http://www.apollosaturn.com

Jean Schneider's encyclopedia of extrasolar planets: http://cfa-www.harvard.edu/planets/f-encycl.html

Darwin exoplanetary mission: http://ast.star.rl.ac.uk/darwin

Corot exoplanetary mission: http://www.obspm.fr/encycl/corot.html

The latest pictures from the Hubble Space Telescope: http://oposite.stsci.edu/pubinfo/pictures.html

Live images of the Sun from the SOHO probe: http://sohowww.nascom.nasa.gov/data/realtime-images.html

PHOTOGRAPHIC ACKNOWLEDGEMENTS

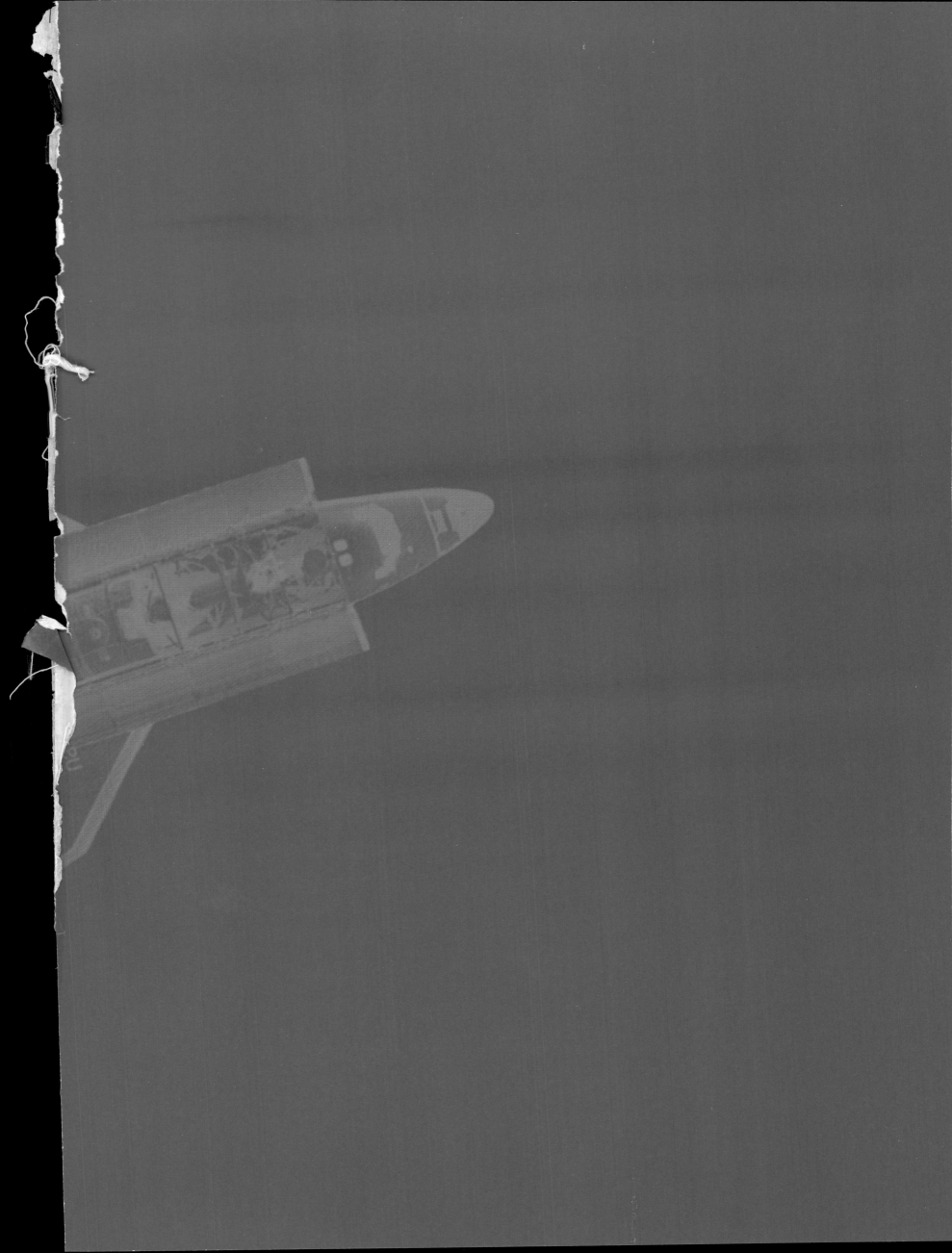

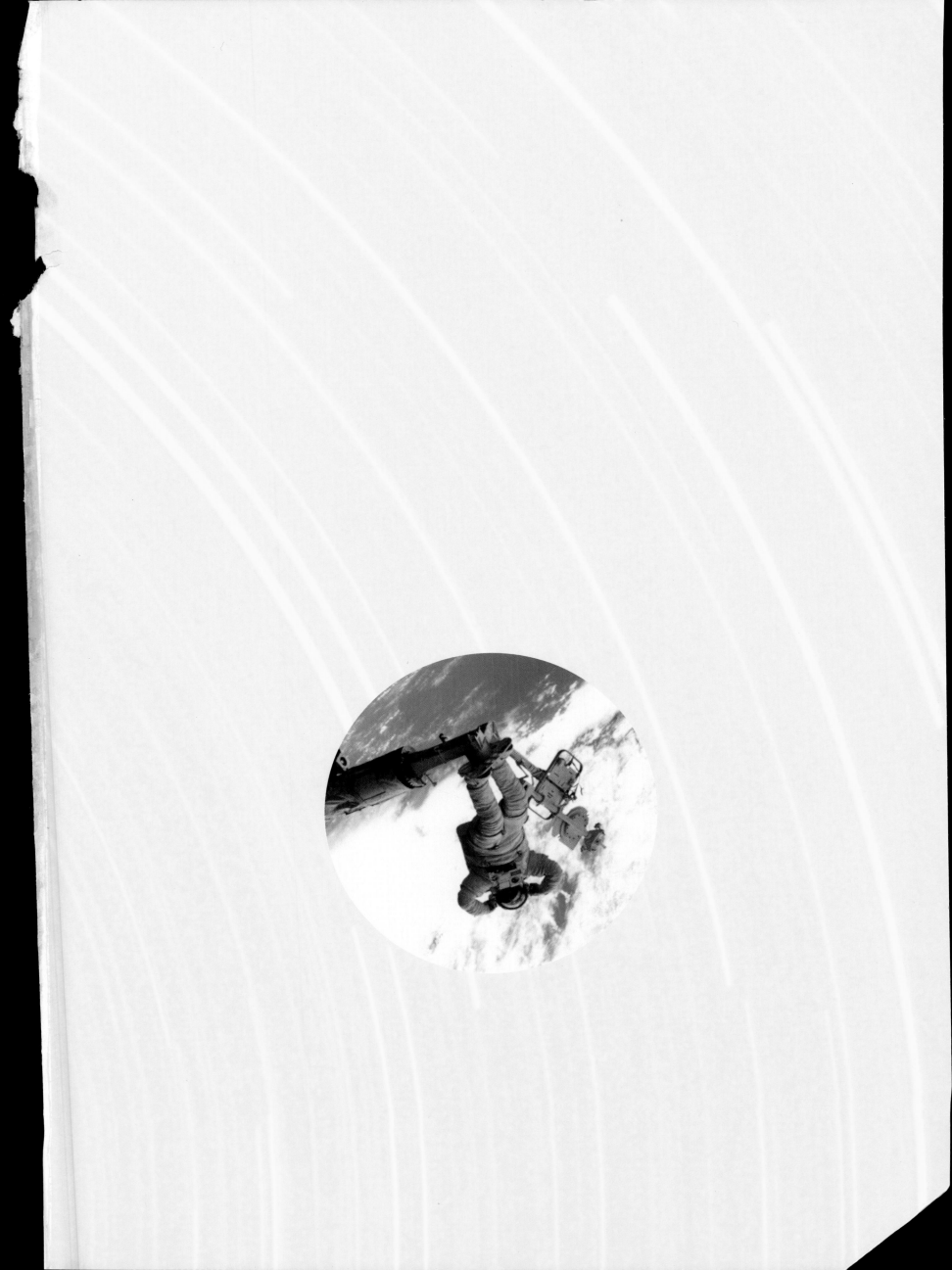

and the general public alike, and it should not be forgotten that, at the end of the day, the conquest of the Solar System is actually financed by the ordinary citizen. Towards the end of the 1990s, the Americans made a number of attempts to visit the Red Planet, suffering several misfortunes: Mars Observer, Mars Polar Lander and Mars Climate Orbiter were all mysteriously lost. It cannot be said that Solar System exploration has been reduced to a routine exercise, like air travel within our own atmosphere. The first of two exceptions is Mars Global Surveyor, launched on 7 November 1996 and going into Mars orbit on 12 September 1997. This orbiting probe is still operating, and has sent back much science data concerning climatic conditions, surface morphology, gravity and magnetic field, among other things. It can also act as a relay for future Mars survey missions. But the most resounding success in this otherwise disastrous decade, during which the Russians also lost a Mars probe (Mars 96) as it descended through the Martian atmosphere, was the small and discreet Mars Pathfinder. At a cost of only 250 million euros, the Americans managed to softland a probe equipped with a six-wheeled robot, able to move around in the rugged Martian environment. This exploit was more than a mere technological leap. In fact, the Russians had already achieved the same on the Moon back in 1970, albeit 500 times closer than Mars. The Mars Pathfinder mission was to spark off a miniature philosophical revolution, as we shall soon see.

US and European probes will soon return to Mars. In 2003 the European probe Mars Express will softland on the icy desert of the Red Planet. Then in 2007 or 2009 the French–American probe Mars Sample Return, launched by Ariane 5, will set off for Mars with the specific aim of collecting Martian rocks and sand and then bringing them back to Earth, sometime between 2010 and 2012.

The main reason for collecting Martian samples is for geological study of a planet that in many ways resembles the Earth. The collection and return of cometary samples, also scheduled for the beginning of the twenty-first century, has a rather different objective, namely the physicochemical study of objects that have very likely remained unchanged since the beginning of the Solar System. This should lead to an improved understanding of the birth and evolution of our Solar System over the past 4.5 billion years. As mentioned above, a European probe Giotto has already taken close-up photos of comet Halley, back in 1986. Various missions now under development on either side of the Atlantic promise great discoveries and images worthy of the best science-fiction films. The American Stardust probe, for instance, which left Earth on 6 February 1999, will reach comet Wild 2 in January 2004. If all goes well, the probe will collect several micrograms of cometary particles and bring them back to Earth in a tiny canister in January 2006. The Contour probe will approach comet Encke in 2003, comet Schwassmann–Wachmann in 2006 and comet Arrest in 2008, photographing them from every angle. The last American mission for the first decade of the twenty-first century will be the probe Deep Space 4, due for launch in 2003. This tiny probe will reach comet Tempel 1 some time in 2005, dropping the Champollion module onto its surface. The latter will bring cometary ice and dust back to Earth in the year 2010.

Following the success of their spectacular Giotto mission, the first to fly by a comet, Europeans have been working on a further probe called Rosetta which will fly aboard the powerful Ariane 5 rocket in January 2003. Its target will be comet Wirtanen. After eight years in space, it will go into orbit around the cometary nucleus in order to accompany it along its orbit for about a year and a half. Then in 2012 Rosetta will drop a small module onto the surface of the comet in order to explore the little icy body and take samples which it will analyse *in situ* using a whole range of highly sophisticated instruments.

Another major objective for planetary scientists will be Saturn's giant moon Titan. This object remains one of the most tantalising in the Solar System. It has only been overflown once, and then for just a few hours, when Voyager 1 flew past some twenty years ago. Unfortunately, Voyager's cameras proved unable to penetrate the thick cloud layers that shroud the surface. From this moment, astronomers have dreamt of studying the little 'hibernating planet', whose atmosphere, like our own, is composed essentially of nitrogen. The honour will fall to the European probe Huygens, specially designed to reveal what is hidden beneath Titan's cloud formations. It was launched aboard the American probe Cassini in October 1997 and will reach Titan in November 2004. Cassini will drop it down through Titan's thick atmosphere, before itself continuing with a systematic four-year exploration of Saturn. Swinging gently from its parachute, Huygens will have about two and a half hours to analyse Titan's atmosphere. It will then softland on the surface and photograph a landscape bathed in permanent twilight. Despite twenty years of careful observation, nobody has yet been able to determine whether Titan's surface is solid or liquid. Huygens has thus been designed to float, should the need arise, on a lake or sea of methane, such as some scientists have predicted, here and there, in the icy desert of this strange and remote planet. Never in the history of Solar System exploration will a probe have softlanded so far away, at an astonishing 1.5 billion kilometres from Earth.

Much further still than Saturn, the planetary pair composed of Pluto and its moon Charon has yet to receive the visit of humanity's space-touring robots. At five billion kilometres from Sun and Earth, these two ice-bound objects constitute the main objective of an ambitious American space mission called the Pluto–Kuiper Express, scheduled to leave Earth between 2004 and 2006 and arrive some time around 2012, before vanishing beyond the distant confines of the Solar System.

But perhaps the most intriguing space mission is one NASA sent out to the giant planet Jupiter. When the Voyager probes discovered the diversity of little worlds gravitating around Jupiter some two decades ago, the Galileo mission was set up, arriving in the Jovian neighbourhood in 1995. This mission has long outlived its original timetable and is still exploring Jupiter's moons in 2001. Among other things, it has sent back amazingly detailed photographs of the surfaces of Jupiter's four largest moons, Io, Europa, Ganymede and Callisto. Galileo's most significant discovery concerns the surface of Europa, which turns out to be a vast ice floe. It very likely conceals beneath it an immense ocean, some 100 kilometres deep and three times the volume of all the seas on our own planet.

A VAST OCEAN UNDER EUROPA'S ICE BANKS

It remains to prove with certainty that such an ocean really does exist, to determine the thickness of the ice layer that protects it, and of course to explore its mysterious depths. Fascinated by this unique world, American planetary scientists have designed the Europa Orbiter mission, which NASA expects to launch in 2006. Three years later, when it arrives in the Jovian system, the probe will go into orbit around Europa. From this vantage point, it will produce accurate maps and measure the ice thickness using radar. If, as is hoped by the Europa Orbiter team, the ice is no more than a few kilometres thick, or even just one kilometre thick, in certain places, the Europa exploration programme will go into its most spectacular phase; NASA is planning to softland hybrid probes on the icy surface, some time around 2020. These will be capable of drilling deeply through the ice until they reach water, then analysing the chemical composition of the biggest ocean in the Solar System. In their wildest dreams, scientists hope they may discover some form of life, formed four billion years ago in the shelter of the icy matrix. Should such a discovery actually be made, planetary exploration will very likely witness a sudden expansion, involving all the planets of the Solar System.

But what will happen in the opposite case? We have already seen that human beings themselves are notable by their absence in the exploration of the Solar System. The planets have been sought out and scrutinised by ever more 'intelligent' robots. Galileo is a case in point. Earth-based engineers have even been able to increase its transmission capabilities by remote intervention. Considering the enormous expense involved in basing men and women aboard the International Space Station in its low Earth orbit, can we seriously envisage sending them further afield, to the Moon and Mars? This was indeed what many once believed. No sooner had the Apollo pioneers returned to Earth in the 1970s than

■ THE HUBBLE SPACE TELESCOPE HAS RECEIVED REGULAR VISITS BY ASTRONAUTS SINCE ITS LAUNCH IN 1990. ITS OBSERVATIONS OF URANUS ARE BETTER THAN THOSE CARRIED OUT BY VOYAGER 2 IN 1986.

futurologists were predicting a human presence on Mars for the 1980s! And yet the dream has gently slipped away. From the 1990s to begin with, it was gradually pushed back to the first years of the new millennium, then 2010, until today, those who advocate a genuine human presence in space, in addition to unmanned space probes, stubbornly picture the fiftieth anniversary celebrations of the historic Apollo 11 flight on the Red Planet in July 2019. To many observers, however, this is a utopian scheme. For one thing, such a project remains technologically unrealistic. Engineers who have been producing scenarios for manned flights to Mars over the past thirty years have found themselves confronted with technical and human constraints that go far beyond those facing the Soviets and Americans when they were engaged upon lunar conquest. Particularly on the human level, both physiological and psychological demands are of a different order. To begin with, the distance and corresponding journey time are roughly 500 times greater. To complicate matters, the distance from Earth to Mars is variable because they follow their orbits independently around the Sun. This means that only certain periods are propitious for launch, the so-called launch windows, both for launch from Earth and return from Mars. In fact it is only possible to go out to or return from Mars when the two planets are in the neighbourhood of their closest approach, which means once every two years or so. Since six months are required for the journey, the crew would have to remain at least one and a half years on the surface of the planet. Adding six months for each of the outward and return journeys, the constraints imposed by celestial mechanics and the relatively small speeds of current spacecraft would thus require a total mission time of two and a half years in space. This should be compared with twelve days for the longest of the Apollo missions. In the spartan conditions of space travel, two and a half years would seem an eternity. In most of the scenarios considered by NASA, the crew would consist of six men and women. They would live in isolation in modules barely larger than those that made up Mir, for both the flights to and from Mars and the long stay on the surface of the Red Planet. Who could stand such confinement? Advocates of a manned mission to Mars often argue that the Russian cosmonaut and doctor Valeri Polyakov lived aboard Mir for more than a year without any difficulty. This may be so, but apart from the fact that a Mars mission would last twice as long, the situation faced by these future Mars explorers would bear no comparison to that endured by Mir's inhabitants. When feeling

■ During the summer of 1997, the Mars Pathfinder probe, a genuine interplanetary photographic reporter, produced sensational images of the surface of the Red Planet, transmitting them live around the world via the Internet.

down, the latter would press a melancholy face against the porthole and contemplate their home planet, that marvellous life-containing bubble, so familiar and friendly, and above all, so close. Indeed, the Mir cosmonauts knew that, in the event of a technical or medical problem, they had only to climb into their Soyuz re-entry craft and return to Earth safe and sound a few hours later. But for the Mars expedition, the crew would leave with no hope whatsoever of returning in the short term, even in the event of damage, accident or illness. If something went wrong on the outward flight, as happened with Apollo 13, the Mars spacecraft would be condemned to a trip lasting at least one year, whatever happened, since it would have to fly out to the Red Planet, revolve about it and then make the long return journey.

WILL ASTRONAUTS EVER GO TO MARS?

Which space agency, which country, which ideology will seek to submit its astronauts to such extraordinary risks? What price will be paid for leaving Earth for the icy deserts of Mars? Added to these problems, quite unprecedented in the history of human exploration, severe technical constraints remain to be fulfilled. A heavy-duty launch vehicle, at least as powerful as Saturn V or Energia will be needed to send the components of the Mars spacecraft into Earth orbit. This implies a new rocket, only as yet existing on paper. The next problem is the safety of the astronauts during the expedition. Unlike the inhabitants of Mir, protected by the Earth's magnetic field, Mars astronauts will be continually subjected to bombardment by solar radiation and cosmic rays arriving from remote stellar explosions. This flux of high-energy particles and radiation is known to be extremely harmful and probably lethal at a high enough dose. No one yet knows exactly how to protect astronauts from this threat during the flights and the stay on Mars. In addition, the flight to Mars and energy production on the planet itself will require the use of nuclear reactors that do not exist today, apart from the fact that such means are bitterly contested by environmental activists who fear the consequences of radioactive fallout in the event of disaster at blast-off. In short, the trip to Mars would involve a technological mobilisation on a par with the Apollo programme, at a cost of between 100 and 500 billion euros which no one could seriously envisage today.

At the same time, behind the dream of conquest, one particular question always lurks in the background: what is the point of

going to Mars? Humans have already made the symbolic step into another world, on 21 July 1969. Why should we repeat the same exploit a bit further out into the Solar System? Especially when Mars Pathfinder clearly demonstrated in 1997 that robots were equal to the task of exploring another planet, moving across the surface, analysing rock samples *in situ*, and photographing the landscape, and all at a far lower cost than any manned flight. This spectacular unmanned mission cost only half as much as a single shuttle flight! Added to this, the failure of Mars Observer, Mars Polar Lander and Mars Climate Orbiter serve as a hard lesson to engineers that space travel remains an extremely high-risk activity, even at the beginning of the twenty-first century. The Viking and Mars Pathfinder missions are witness to the fact that even virtual visits through unmanned probes can offer a dizzying perspective of the Red Planet whilst human observers remain within the safe embrace of their own planet.

So when will human beings first set foot on Mars? No one knows. Such an expedition might become relevant if Mars-based robots were to find traces of life there, either living or fossilised. For the past forty years, groups of scientists associated with these Mars missions have always used the discovery of life to justify costs in the eyes of the general public. Yet the Viking probes, specifically designed to seek out life, found no trace whatever in the icy Utopia and Chryse plains. Again, a trip to Mars might become a more realistic prospect if the travel time could be drastically reduced. But for the moment, we are a long way from such a technological breakthrough. There are no ideas on the table

■ IS IT WORTH RISKING THE LIVES OF ASTRONAUTS BY SENDING THEM TO MARS? IN THE YEARS TO COME, THE ARTIFICIAL INTELLIGENCE OF SPACEBORNE ROBOTS WILL ALLOW THEM TO SURPASS HUMAN CAPABILITIES IN THE HARSH CONDITIONS BEYOND OUR PLANET.

that might reduce the round trip to, say, one or two months. Will this be possible in 2030? Or will we set off for Mars on the first centenary of the Apollo 11 flight, in July 2069? The only way to gauge the prospects for humankind in space is to follow the development of the International Space Station, which has been revolving noiselessly about the Earth since 20 November 1998. For it is here that the future of human activities in space will be decided over the years and decades to come. For the time being, it is not known whether the countries that have subscribed to this huge project will be able to carry it to completion and whether the deadlines and budget will be respected. It is not known whether the scientists carried aboard will make fundamental discoveries, nor whether new technoscientific applications will at last be developed in weightless conditions. When the International Space Station is completed, and the industry which supports NASA goes in search of public financing, will the American space agency be forced to advocate a voyage to Mars as the only end which can justify its manned flights, and even its very existence? Or will a new and wealthy world power reveal itself before then, with dreams of asserting itself in the political, scientific and economic arenas through feats that rival the Apollo programme? Could China, India or Brazil be the great conquering nation of the future?

Today, all these questions remain open. Humanity's future in space hangs upon the answers that will be brought to them over the next twenty years. And only then will we know whether, one day, astronauts aboard some proud celestial spaceship will once again experience the giddying adventure of escape from Earth.

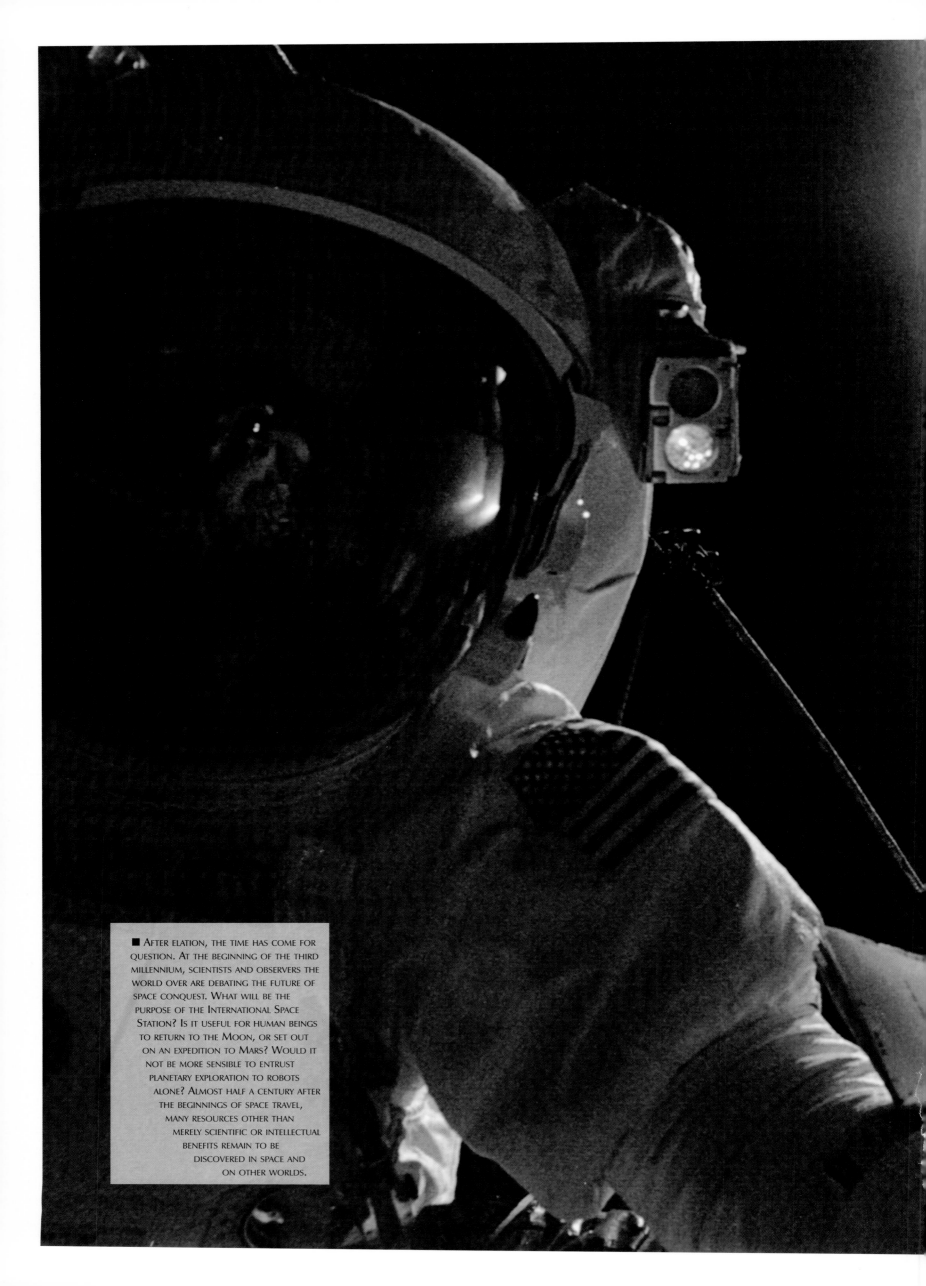

■ AFTER ELATION, THE TIME HAS COME FOR
QUESTION. AT THE BEGINNING OF THE THIRD
MILLENNIUM, SCIENTISTS AND OBSERVERS THE
WORLD OVER ARE DEBATING THE FUTURE OF
SPACE CONQUEST. WHAT WILL BE THE
PURPOSE OF THE INTERNATIONAL SPACE
STATION? IS IT USEFUL FOR HUMAN BEINGS
TO RETURN TO THE MOON, OR SET OUT
ON AN EXPEDITION TO MARS? WOULD IT
NOT BE MORE SENSIBLE TO ENTRUST
PLANETARY EXPLORATION TO ROBOTS
ALONE? ALMOST HALF A CENTURY AFTER
THE BEGINNINGS OF SPACE TRAVEL,
MANY RESOURCES OTHER THAN
MERELY SCIENTIFIC OR INTELLECTUAL
BENEFITS REMAIN TO BE
DISCOVERED IN SPACE AND
ON OTHER WORLDS.

Voyage without motion

■ Is the sky too vast for human ambition? The Milky Way, our galaxy, contains close on a thousand billion stars. This means that there are a hundred times more stars in our little corner of the cosmos than there are human beings on Earth. Science-fiction writers have long accustomed us to the idea of our species spreading across and conquering the whole Universe. But this future galactic empire will never be more than a dream.

■ Viewed from Earth with the naked eye, the Red Planet does indeed seem remote. A round trip to Mars of the kind that astronauts have been dreaming of in the coming decades would take more than two years.

At the beginning of the third millennium, astronauts once again seek to determine their destiny in space. The days of the pioneers struggling to control restive and highly-strung spacecraft are gone. The amazing lunar odyssey has been successfully brought to completion and humankind's apprenticeship to life in space accomplished with the help of the wonderful Mir space station. So the time has come to ask what will be the future role of astronauts in space. The International Space Station only serves to trace out the question mark in the heavens. Even the most enthusiastic supporters of space exploration contest the usefulness of such an exploit, beyond its technological and geopolitical dimensions. The severest critics even go so far as to suggest that its sole purpose is to justify the existence of the American space shuttle!

Half a century after testing the first space-bound rockets, we are compelled to observe that, apart from the extraordinary commercial exploitation of space which has brought success to hundreds of telecommunications and Earth-observation satellites, and excepting the wonderful journeys of space probes sent out to the four corners of the Solar System, space exploration is now biding its time. The role of human beings in the cosmos is currently in question, along with all our dreams of exploration so present throughout the last century, as expressed by science-fiction writers and visionary engineers. All have pulled up short before the hard face of reality. The fact is that it is difficult to travel in space, and it costs a lot. The staggering progress made in the aeronautic industry, which took only sixty years to completely transform planetwide communications by producing aircraft capable of transporting millions of people around the world each day, has had no echo in the domain of space travel. In 1968, Stanley Kubrick's marvellous film *2001: A space odyssey* owed a large part of its success to the fact that many viewers at the time took the story at face value. They could easily imagine that in 2001 men and women would be living in space, that they would be setting off for the Moon and Mars. It is an ironic twist of fate that it is precisely now, at such a symbolic moment when we are beginning a new millennium, that human presence in space should reach such a low point. We may ask whether the astronauts' return to Earth will remain a permanent feature, or whether the slump that we are experiencing at present is merely a prelude to a much greater wave of conquest.

The present state of astronautics gives little reason for optimism. There is an obvious gulf between the celestial dreams of cosmonauts and the lack of any long-term view amongst

■ JUPITER MOVES SLOWLY
THROUGH THE CONSTELLATION OF
CAPRICORNUS. THE TWENTIETH
CENTURY WILL GO DOWN IN
HISTORY AS A PRIVILEGED EPOCH,
A PRECIOUS MOMENT WHEN
HUMANKIND MANAGED TO
BREAK ITS EARTHLY BONDS. IN
FEWER THAN THIRTY YEARS THE
WHOLE OF THE SOLAR
SYSTEM HAS BEEN EXPLORED
BY SPACEBORNE ROBOTS.

political leaders. This gulf is only widened by the inadequate state of today's technoscientific capacities. We hear of vast science parks in space, exploitation of inexhaustible mineral resources on asteroids, colonisation of the Moon and Mars, or even the whole Solar System, before humanity finally makes its inevitable leap towards the stars. Among these futuristic accounts, how may we distinguish the probable or even the possible from the irrational and the utopian? All these projects take it for granted that access to space will soon be as simple and economical as flight within the Earth's atmosphere, which makes it possible for us travel from one side of the Earth to the other at will.

But this is to forget that aeronautics and astronautics are quite different disciplines. An aircraft escapes from the Earth's surface, not from its gravitational field. It flies because it succeeds in resting its weight upon the air like a bird. Unfortunately, a bird cannot fly out to the stars. Since Einstein discovered almost nine decades ago that gravity is not a force but rather a manifestation of the curvature of space–time, we must get accustomed to imagining the Earth's surface as the bottom of a deep gravitational well. Escaping from the Earth's gravitational attraction amounts in some sense to climbing the smooth, vertical, transparent walls of this strange spatio-temporal chasm. In consequence, despite its resemblance to a DC 3, the American shuttle is not an aircraft. The tremendous feat of wrenching seven astronauts and a few tonnes of cargo from the Earth's grasp and raising them to an altitude of a mere 500 kilometres requires the combined power of sixty Airbuses.

So what of the Moon, Mars and the asteroids? Today, it costs more than 20 000 euros to place a payload of one kilo into Earth orbit using the shuttle. This is a long way from the ten euros per kilo typical for airborne cargos. Using our current space capabilities, it would cost about 100 million euros to return one kilo of Moon rock, whilst one billion euros would be required per kilo of Martian rock samples. We may well wonder what extraordinary technological revolution is going to make it feasible to exploit the mineral resources of the lunar and Martian soils, or operate mines on remote asteroids. As a corollary: what wonderful mineral deposits must these heavenly bodies possess – in addition to the treasury of mystery and beauty concealed by their very remoteness – to make their exploitation economically viable?

■ SILENT AND INVISIBLE, HUMANITY'S FOUR AMBASSADORS SLIP EFFORTLESSLY ACROSS THE VAST EXPANSE OF SPACE. BEFORE THEM, FOR

ROBOTS TO THE RESCUE

Once again, it seems that the real future of space conquest must begin with unmanned flight, artificial satellites, probes and robots. In his book *Une autre histoire de l'espace*, French author Alain Dupas invoked the creation of an Internet on the scale of the whole Solar System. The idea behind this 'interplanetary village' is all the more realistic and exciting in that it has begun to see the light of day over the past few years. In order to understand how planets evolve, and in particular, how their complex atmospheres function, scientists need to maintain a long-term presence. A whole armada of meteorological satellites has been required over the past three decades to gain even an imperfect handle on the Earth's weather systems. In the twenty or thirty years to come, Venus, Mars and Jupiter will be equipped with their own networks of meteorological and communications satellites. These will allow not only specialists, but also the general public to follow climatic change on these planets in real time. The European Space Agency has proven since 1995 that such virtual exploration is indeed possible. Its solar meteorology satellite SOHO, positioned 1.5 million kilometres from Earth, transmits spectacular live images of raging solar storms several times a day. And in July 1997, NASA accomplished the amazing feat of transmitting live landscape panoramas from the Red Planet to the screens of more

than 500 million personal computers, together with a daily Martian weather bulletin. It seems clear that the future of space conquest must lie in the production of images. These serve not only to convey scientifically useful information but also to fuel the enthusiasm of the general public in the hope that financial support will always be forthcoming. American research teams have learnt this lesson. Almost no scientific mission is now designed without its resident TV camera.

Is it conceivable that private enterprise might one day turn astronomical imaging into a business? Why not? The Mir space station was often used as an advertising studio and was at one point coveted by a large American firm with a view to producing a full-length film featuring international film stars. But quite apart from the fact that the actors in question had no desire to risk their lives in orbit, or to suffer the unpleasant experience of space sickness, the undertaking proved too complex and too expensive. In some respects, however, the idea had made its way. It will soon be possible to send a small probe to the Moon equipped with a wide-field, high-resolution colour camera, thanks to advances made in miniaturised optoelectronics, at a cost not exceeding that of a Hollywood superproduction. We may then be able to dream of a stroll on the Moon or even Mars, a kind of hybrid mission combining scientific interest with space tourism, aboard space vehicles touring around the most beautiful extraterrestrial scenery.

Let us imagine some time in the near future, between 2015 and 2020, for example. In planetariums, IMAX® domes or even in separate virtual reality cubicles connected to the Internet, people around the world will be able to witness in real time the exploration of another world by mobile robots dispatched to the four corners of the Martian desert. Day by day, the astonishing scenery of Mangala Vallis or the canyons of the Tharsis Rise will be gradually revealed on our screens: clear early mornings in the heart of winter when the salmon pink sky turns to dark blue and the hoar frost forms on orange-tinged rocks at temperatures of $-128°C$; lacklustre days when storms are brewing and the Sun lies hidden behind a haze, the gusting wind whipping up desert dusts to veil the distant hills; or summer afternoons when the Sun heats up the surface and the temperature climbs and climbs, as if by miracle, to reach $0°C$, or even $+10°C$. The probe then parks itself for a few days in order to observe the way the light varies across the landscape. Methodically, it photographs scenes of extraordinary natural beauty. On each of these mobile robots, a microphone records the gentle lamentation of the wind or the sound of the red sand grating under its metal wheels. As the months go by, the exploration proceeds. Like a giant, untiring insect, another probe cautiously follows the track of an ancient Martian water course where it was so long hoped that fossil life forms would be detected. Yet another reaches the edge of a tributary canyon to the great Valles Marineris in order to film one of the most extensive panoramas in the Solar System. The camera pans up the 9000 metres of the canyon's monumental walls, or monitors the gradual progress of clouds gathering in the depths of the gorge, recording the infinite and subtle variations of pink and orange as the Sun sets on planet Mars.

At the end of the last century, the Mars Pathfinder mission gave us a foretaste of what would be possible tomorrow, or even today, at the very beginning of the twenty-first century, without the need for extraordinary technological breakthroughs or prohibitive spending.

As far as human beings are concerned, we have already seen that any effective human presence in space will depend above all else on what we are able to do with the International Space Station between now and 2015 or 2020, and on the projects that supersede it. But there is a further crucial factor when we consider the future of manned space flight, namely, the development of

launch vehicles. Today, space can be reached using two types of launch vehicle: first there is the conventional rocket, simple and powerful but costly because it cannot be reused; then there is the shuttle, able to make many round trips between Earth and space, but so exorbitantly expensive to maintain that a single launch effectively amounts to 350 million euros. Paradoxically, the shuttle works out more expensive than a conventional rocket!

The Semiorka rocket and the American space shuttle are destined to meet space transportation needs for a long time to come. Indeed, NASA intends to use its fleet of four shuttles for another fifteen or twenty years, that is, until the end of the nominal lifetime of the ISS. Towards 2020 or 2030, engineers hope to see a new generation of launch vehicles. These will be reusable like the shuttle, in manned or unmanned form. Lighter, more powerful, with enhanced aerodynamics, they may even be able to take off from a horizontal position like an ordinary aircraft. They would not require such costly repairs between flights as current shuttles. The only hitch is that such genuine space planes will require the invention of extremely powerful and sophisticated engines, no easy task for today's engineers. Their dream is to design a hybrid engine able to operate first in the Earth's atmosphere where oxygen is available, then in space where it would consume liquid oxygen carried aboard. Such an engine could make tremendous savings on propellant and would allow the launcher to take off without boosters or hefty propellant tanks.

If this kind of engine is ever developed and if it does indeed cut the cost of space transportation by a factor of ten or fifty as engineers are hoping, space conquest may be able

■ 'THE ETERNAL SILENCE OF THESE INFINITE REACHES TERRIFIES ME,' ADMITTED FRENCH PHILOSOPHER BLAISE PASCAL AS HE CONTEMPLATED THE HEAVENLY VAULT IN THE SEVENTEENTH CENTURY. FOR THE LAST FORTY

to add the inexhaustible and unexpected resource of tourism to the still hypothetical technoscientific applications. Even if deserted by science and industry, Earth orbit may prove to be a wonderful play area for a new species of holidaymaker, in search of excitement and wealthy enough to foot the bill. This kind of holiday already exists on Earth. Many tour operators invite amateur explorers to discover the Antarctic or the high peaks of the Himalayas. Those who have spent a night at an altitude of 8000 metres on the southern flank of Mount Everest have surely procured a foretaste of other-world landscapes. Today, such expeditions cost in the region of 20 000 to 30 000 euros. In 2030 or 2040, this could well be the price to pay for the first touristic escapades in space, until orbiting palaces are developed in the mid-century providing a room with a view over the seven seas.

What then? We would need to see a further technological leap accompanied by another drop in the cost of space transportation, by a factor of ten or hundred, if the next (still hypothetical) phase in the occupation of space were ever to become feasible: the exploitation of lunar helium 3. This rare isotope of helium does not exist on Earth. However, it is produced in large quantities by the Sun and has been accumulating in the lunar regolith for billions of years. A century from now, it may be possible to collect it and return it to Earth. This almost perfect fuel, with high energy yields

and no harmful by-products, could supply nuclear fusion reactors. But apart from the fact that such reactors are unlikely to exist for several decades, if they are ever built at all, the future of terrestrial energy supplies may well lie with wind and solar power stations, thus banishing the atom from this domain of human technology. A return to the Moon would then be postponed. We would have to resign ourselves to the inevitable, despite the irresistible appeal of the mysterious landscapes we have glimpsed on the Moon, Mars, Europa and Titan. In this case we would have no practical reason to return for a long time to come. Further visits to such worlds would no longer concern our generation, nor those of our children and grandchildren. And yet, as the decades go by, future waves of space probes will make the planetary surfaces in our Solar System as familiar to us as the surface of our own planet. Soon the whole of humankind should become aware that it belongs to a much vaster world than its fragile blue planet. Paradoxically, through accumulated knowledge of all worlds making up the Solar System, scientific effort will not aim to colonise those worlds, an unrealistic project for at least the next few centuries, but rather to restore to our own planet its former ecological equilibrium. The result may be that we render the Earth more hospitable for humanity.

Finally, once all the planets in the Solar System have been visited, explored, studied, and equipped with their own global communications network, human beings will be able to tackle what is likely to be the greatest adventure of the new millennium. For there is a new feature today which no futurologist, no science-fiction writer, no scientist, and no astronaut had predicted during the past few decades, dedicated as they were to the Moon and Mars and other bodies quite close to home. This is the discovery, at the turn of the millennium, of dozens of other planets further afield, a discovery which has somewhat eclipsed the few worlds that revolve around our own tiny Solar System. And the numbers of these exoplanets is increasing rapidly, so that we may soon expect to know of hundreds, then thousands and later even millions of other worlds.

TEN BILLION OTHER WORLDS

It all started in 1995 at the Haute-Provence observatory in the south of France, when two Swiss astronomers, Michel Mayor and Didier Queloz discovered the very first exoplanet. It was a strange object, slightly less massive than Jupiter and gravitating around 51 Pegasus, a star that resembles our own. But the new planet in the constellation of Pegasus was only the first messenger bringing news of myriad other such objects just waiting to reveal themselves to new detection methods. At the end of 2000, more than sixty exoplanets had already been located, gravitating around stars within a few dozen light years from the Sun. In other words, they already outnumbered the planets of our own Solar System together with all their natural satellites. For the time being it is

159

impossible to see, or even to photograph these remote planets. Swamped by the light from their accompanying star, they are effectively invisible and can only be detected indirectly. Astronomers reveal their presence by observing perturbations they induce in the motion of their star, or else through periodic eclipses as they pass in front of it. Today only the very biggest planets can be spotted in this way. Most have a mass between those of Saturn and Jupiter. Some are almost close enough to touch their star, whilst others orbit at the same kind of distance as Mercury, Venus, Earth or Mars in relation to the Sun.

With these extraordinary discoveries, whose philosophical significance no one has yet been able to gauge, we are perhaps on the brink of a new scientific paradigm that could long trouble its sister disciplines of astrophysics and astronautics. For these objects, still remote, mysterious and abstract to most of us, will gradually become as real as the planets in our own Solar System. Over the next twenty years, as astronomical observation strides ahead, the exoplanets will eventually be photographed. Their mass, size, temperature, and even chemical composition will be opened up to investigation. But above all, small planets comparable in size with the Earth, Venus and Mars will finally be brought to light.

What a marvellous field for discovery and creativity, cutting across all aspects of human culture! These remote worlds, almost infinite in number, with their astonishingly varied landscapes and mind-blowing novelty, will call out to us as surely as the Moon and Mars once attracted all those men and women fascinated by the spectacle of a starry night.

TRAVEL ACROSS THE UNIVERSE WILL BE VIRTUAL

Scientists and engineers involved in space exploration will inevitably turn their attention to these new worlds. In fact the powerful telescopes designed to discover and study exoplanets can only be set up in the void of space. In 2004 the French satellite Corot will go into orbit to observe tens of thousands of stars in the hope of detecting eclipses caused by several dozen planets the size of Earth. Then in 2009, Hubble's successor the Next Generation Space Telescope will observe the most massive and hottest exoplanets in the infrared before being joined by the European probe GAIA, expected to detect tens of thousands of celestial objects the size of Jupiter!

But already, another old question raises itself. Both NASA and

■ THE ANDROMEDA GALAXY, TWIN SISTER OF THE MILKY WAY, CONTAINS MORE THAN A THOUSAND BILLION STARS. IN ORDER TO REACH US FROM THIS DISTANCE, LIGHT TAKES